The Science of Science

This is the first comprehensive overview of the "science of science," an emerging interdisciplinary field that relies on big data to unveil the reproducible patterns that govern individual scientific careers and the workings of science. It explores the roots of scientific impact, the role of productivity and creativity, when and what kind of collaborations are effective, the impact of failure and success in a scientific career, and what metrics can tell us about the fundamental workings of science. The book relies on data to draw actionable insights, which can be applied by individuals to further their career or decision makers to enhance the role of science in society. With anecdotes and detailed, easy-to-follow explanations of the research, this book is accessible to all scientists and graduate students, policymakers, and administrators with an interest in the wider scientific enterprise.

Dashun Wang is an Associate Professor at Kellogg School of Management, and (by courtesy) McCormick School of Engineering, Northwestern University, where he directs the Center for Science of Science and Innovation. He is a recipient of the AFOSR Young Investigator Award and was named Poets&Quants Best 40 Under 40 Business School Professors in 2019.

Albert-László Barabási is Robert Gray Dodge Professor of Network Science and Distinguished University Professor at Northeastern University, where he directs the Center for Complex Network Research. He holds appointments in the Departments of Physics and the College of Computer and Information Science, in the Department of Medicine at Harvard Medical School, at Brigham and Women's Hospital in the Channing Division of Network Science, and the Department of Network and Data Science at Central European University in Budapest, Hungary.

The Science of Science

Dashun Wang
Northwestern University, Evanston, Illinois

Albert-László Barabási
Northeastern University, Boston

CAMBRIDGE UNIVERSITY PRESS

CAMBRIDGE
UNIVERSITY PRESS

University Printing House, Cambridge CB2 8BS, United Kingdom

One Liberty Plaza, 20th Floor, New York, NY 10006, USA

477 Williamstown Road, Port Melbourne, VIC 3207, Australia

314–321, 3rd Floor, Plot 3, Splendor Forum, Jasola District Centre,
New Delhi – 110025, India

79 Anson Road, #06–04/06, Singapore 079906

Cambridge University Press is part of the University of Cambridge.

It furthers the University's mission by disseminating knowledge in the pursuit of
education, learning, and research at the highest international levels of excellence.

www.cambridge.org
Information on this title: www.cambridge.org/9781108492669
DOI: 10.1017/9781108610834

© Cambridge University Press 2021

This publication is in copyright. Subject to statutory exception
and to the provisions of relevant collective licensing agreements,
no reproduction of any part may take place without the written
permission of Cambridge University Press.

First published 2021

Printed in Singapore by Markono Print Media Pte Ltd

A catalogue record for this publication is available from the British Library.

ISBN 978-1-108-49266-9 Hardback
ISBN 978-1-108-71695-6 Paperback

Cambridge University Press has no responsibility for the persistence or accuracy
of URLs for external or third-party internet websites referred to in this publication
and does not guarantee that any content on such websites is, or will remain,
accurate or appropriate.

To Tian

CONTENTS

Acknowledgements viii

Introduction 1

Part I   The Science of Career   5

1. Productivity of a Scientist   7
2. The *h*-Index   17
3. The Matthew Effect   28
4. Age and Scientific Achievement   39
5. Random Impact Rule   51
6. The Q-Factor   60
7. Hot Streaks   71

Part II   The Science of Collaboration   81

8. The Increasing Dominance of Teams in Science   85
9. The Invisible College   95
10. Coauthorship Networks   102
11. Team Assembly   110
12. Small and Large Teams   124
13. Scientific Credit   134
14. Credit Allocation   147

Part III   The Science of Impact   159

15. Big Science   161
16. Citation Disparity   174
17. High-Impact Papers   184
18. Scientific Impact   189
19. The Time Dimension of Science   197
20. Ultimate Impact   209

Part IV   Outlook   221

21. Can Science Be Accelerated?   223
22. Artificial Intelligence   231
23. Bias and Causality in Science   241

Last Thought: *All* the Science of Science   252

Appendix A1 Modeling Team Assembly   254
Appendix A2 Modeling Citations   257
References   270
Index   296

ACKNOWLEDGEMENTS

As teams are the fundamental engine for today's innovation, we must thank our lab members first and foremost. Indeed, our lab members contributed to this project in some of the most fundamental ways – not only did their work compose a large part of the initial book proposal, it also contributed fresh insights which grew into several new chapters, broadening and deepening our understanding of the subjects we set out to explore. These team members include Roberta Sinatra, Chaoming Song, Lu Liu, Yian Yin, Yang Wang, Ching Jin, Hua-Wei Shen, Pierre Deville, Michael Szell, Tao Jia, Lingfei Wu, Zhongyang He, Jichao Li, Binglu Wang, Suman Kalyan Maity, Woo Seong Jo, Jian Gao, Nima Dehmamy, Yi Bu, David Moser, Alex Gates, Junming Huang, Qing Ke, and Xindi Wang. Our science of science journey has been infinitely more fun because we had the opportunity to work and learn in such inspiring company.

Chapter 9 on the invisible college discusses the "bright ambience" of brilliant minds, a concept that we experienced vividly while writing this book. Indeed, so many friends and colleagues have been generous with their time and expertise along the way, that listing some of those risks excluding many others. We feel compelled to express our sincere thanks to Luis Amaral, Sinan Aral, Pierre Azoulay, Federico Battiston, Jeanne Brett, Elizabeth Caley, Manuel Cebrian, Damon Centola, Jillian Chown, Noshir Contractor, Ying Ding, Yuxiao Dong, Tina Eliassi-Rad, Eli Finkel, Santo Fortunato, Morgan Frank, Lee Giles, Danny Goroff, Shlomo Havlin, Cesar Hidalgo, Travis Hoppe, Ian B. Hutchins, Ben Jones, Brayden King, Rebecca Meseroll, Sam Molyneux,

Karim Lakhani, David Lazer, Jessie Li, Zhen Lei, Jess Love, Stasa Milojevic, Federico Musciotto, Willie Ocasio, Sandy Pentland, Alex Petersen, Filippo Radicchi, Iyad Rahwan, Lauren Rivera, Matt Salganik, George Santangelo, Iulia Georgescu, Ned Smith, Paula Stephan, Toby Stuart, Boleslaw Szymanski, Arnout van de Rijt, Alessandro Vespignani, John Walsh, Ludo Waltman, Kuansan Wang, Ting Wang, Adam Waytz, Klaus Weber, Stefan Wuchty, Yu Xie, Hyejin Youn.

Among the many people who made this journey possible, our special thanks go to two of them in particular, whose "bright ambience" is especially omnipresent throughout the pages and over the years. Brian Uzzi is not only always generous with his time, he also has remarkable insight. Time and again he manages to take ideas we're struggling to present and effortlessly elevates them. We are also grateful to Brian for championing the somewhat radical idea that physicists can contribute to the social sciences. Our colleague James Evans has been a close friend and collaborator, and several ideas discussed in this book would have not been born without him, including but not limited to the concluding remarks on engaging "all the science of science" for the future development of the field.

We are extremely grateful for the generous research support we have received over the years. In particular Riq Parra from AFOSR has been a true believer from the very beginning, when few people knew what we meant by the "science of science." Many concepts discussed in this book would not have been possible without his strong and continued support. Dashun also wishes to express his special thanks to the Kellogg School of Management, an institution that offered a level of support and trust that most researchers only dream of.

Many unsung heroes contributed to this text, collectively logging countless hours to help guide the book along. We have benefited tremendously from the excellent and dedicated editorial assistance of Carrie Braman, Jake Smith, and James Stanfill, as well as Enikő Jankó, Hannah Kiefer, Alanna Lazarowich, Sheri Gilbert, Michelle Guo, and Krisztina Eleki. Special thanks to Alice Grishchenko who took on the brave redesign effort, diligently redrawing all the figures and giving a visual identity to the book. Yian Yin and Lu Liu have been our go-to people behind the scenes, springing into action whenever we needed help, and contributing to a variety of essential tasks ranging from data analysis to managing references.

Special thanks to our very professional publishing team at Cambridge, from our enthusiastic champion, Simon Capelin, to the editorial team who worked tirelessly to move the project forward and across the finish line: Roisin Munnelly, Nicholas Gibbons, Henry Cockburn, and Marion Moffatt. Thank you for all your help, and for putting up with numerous missed deadlines.

And finally, Dashun would like to thank his wife Tian Shen, and dedicate this book to her. Thank you, for everything.

INTRODUCTION

Scientific revolutions are often driven by the invention of new instruments – the microscope, the telescope, genome sequencing – each of which have radically changed our ability to sense, measure, and reason about the world. The latest instrument at our disposal? A windfall of digital data that traces the entirety of the scientific enterprise, helping us capture its inner workings at a remarkable level of detail and scale. Indeed, scientists today produce millions of research articles, preprints, grant proposals, and patents each year, leaving detailed fingerprints of the work we admire and how they come about. Access to this data is catalyzing the emergence of a new multidisciplinary field, called *science of science*, which, by helping us to understand in a quantitative fashion the evolution of science, has the potential to unlock enormous scientific, technological, and educational value.

The increasing availability of all this data has created an unprecedented opportunity to explore scientific production and reward. Parallel developments in data science, network science, and artificial intelligence offer us powerful tools and techniques to make sense of these millions of data points. Together, they tell a complex yet insightful story about how scientific careers unfold, how collaborations contribute to discovery, and how scientific progress emerges through a combination of multiple interconnected factors. These opportunities – and the challenges that come with them – have fueled the emergence of a new multidisciplinary community of scientists that are united by their goals of understanding science. These practitioners of science of science use the scientific methods to study themselves, examine projects that work as well as those that fail, quantify the patterns that characterize discovery and invention, and offer lessons to improve science as a whole. In this book, we aim to introduce this burgeoning field – its rich historical context, exciting recent developments, and promising future applications.

We had three core audiences in mind as we wrote this book. The primary audience includes any scientist or student curious about the mechanisms that govern our passion, science. One of the founding fathers of the science of science, Thomas Kuhn, a physicist turned philosopher, triggered worldwide interest in the study of science back to 1962 with the publication of *The Structure of Scientific Revolutions*. Kuhn's notion of "paradigm shift" today is used in almost every creative activity, and continues to dominate the way we think about the emergence and acceptance of new ideas in science. In many ways, the science of science represents the next major milestone in this line of thinking, addressing a series of questions that are dear to the heart of every scientist but may well lay outside of the Kuhnian worldview: When do scientists do their best work? What is the life cycle of scientific creativity? Are there signals for when a scientific hit will occur in a career? Which kinds of collaboration triumph and which are destined to for disaster? How can young researchers maximize their odds of success? For any working scientist, this book can be a tool, providing data-driven insight into the inner workings of science, and helping them navigate the institutional and scholarly landscape in order to better their career.

A broader impact of the science of science lies in its implications for policy. Hence, this book may be beneficial to academic administrators, who can use science of science to inform evidence-based decision-making. From department chairs to deans to vice presidents of research, university administrators face important personnel and investment decisions as they try to implement and direct strategic research. While they are often aware of a profusion of empirical evidence on this subject, they lack cohesive summaries that would allow them to extract signals from potential noise. As such, this book may offer the knowledge and the data to help them better take advantage of useful insights the science of science community has to offer. What does an *h*-index of 25 tell us about a physics faculty member seeking tenure? What would the department most benefit from: a junior vs. a senior hire? When should we invest in hiring a superstar, and what can we expect their impact will be?

We also hope that program directors with National Science Foundation (NSF), National Institutes of Health (NIH), and other public and private funding agencies will find the book useful for supporting high-performing individuals and teams to best address science's emerging challenges. Many civilian and military government agencies, nonprofits, and private foundations are already collecting data and developing tools rooted in science of science. The framework offered in the coming chapters will allow them to utilize this data in a way that best serves their own purposes, helping them set up more effective funding mechanisms, and ultimately benefitting both science and society.

The changing landscape of science also affects scholarly publishers, who often compete to publish articles that will impact the direction and the rate of scientific progress. We hope journal editors will also find science of science useful for a range of practical purposes – from understanding the natural life cycle of a discovery's impact to identifying hit ideas before they become hits – which may, in turn, augment the impact of what they publish.

Lastly, this book is intended for scientists who are currently involved in science of science research, or for those who wish to enter this exciting field. It is our aim to offer the first coherent overview of the key ideas that currently capture the discipline's practitioners. Such an overview is necessary, we believe, precisely because our community is highly interdisciplinary. Indeed, key advances in the science of science have been generated by researchers in fields ranging from the information and library sciences to the social, physical, and biological sciences to engineering and design. As such, approaches and perspectives vary, and researchers often publish their results in venues with non-overlapping readership. Consequently, research on the science of science can be fragmented, often along disciplinary boundaries. Such boundaries encourage jargon, parochial terms, and local values. In the book we aim to summarize and translate the insights from highly diverse disciplines, presenting them to students and researchers cohesively and comprehensively. We will not only emphasize the common intellectual heritage of the diverse set of ideas that coexist in the field, but also provide approaches to orient new research. Thus, we hope the book will be a resource for interested students and researchers just discovering the field.

The book is structured in four parts: *The Science of Career* focuses on the career path of individual scientists, asking when we do our best work and what distinguishes us from one another. *The Science of Collaboration* explores the advantages and pitfalls of teamwork, from how to assemble a successful team to who gets the credit for the team's work. *The Science of Impact* explores the fundamental dynamics underlying scientific ideas and their impacts. The *Outlook* part summarizes some of the hottest frontiers, from the role of AI to bias and causality. Each part begins with its own introduction which illuminates the main theme using questions and anecdotes. These questions are then addressed in separate chapters that cover the science relevant to each.

By analyzing large-scale data on the prevailing production and reward systems in science, and identifying universal and domain-specific patterns, science of science not only offers novel insights into the nature of our discipline, it also has the potential to meaningfully improve our work. With a deeper understanding of the precursors of impactful science, it will be possible to develop systems and policies that more reliably improve the odds of success for each scientist and science investment, thus enhancing the prospects of science as a whole.

Part I: THE SCIENCE OF CAREER

Albert Einstein published 248 papers in his lifetime, Charles Darwin 119, Louis Pasteur 172, Michael Faraday 161, Siméon Denis Poisson 158, and Sigmund Freud 330 [1]. Contrast these numbers with the body of work of Peter Higgs, who had published only 25 papers by the age of 84, when he received the Nobel Prize for predicting the Higgs boson. Or think of Gregor Mendel, who secured an enduring legacy with only seven scientific publications to his name [2].

These differences show that in the long run what matters to a career is not productivity, but impact. Indeed, there are remarkable differences among the impact of the publications. Even for star scientists, of all papers they publish, at most a few may be remembered by a later generation of scientists. Indeed, we tend to associate Einstein's name with relativity and Marie Curie with radioactivity, while lacking general awareness of the many other discoveries made by each. In other words, one or at most a few discoveries – the outliers – seem to be what define a scientist's career. So, do these outliers accurately represent a scientific career? Or did these superstar scientists just get lucky in one or a few occasions along their careers?

And, if only one or at most a few papers are remembered, when do scientists make that defining discovery? Einstein once quipped, "A person who has not made his great contribution to science before the age of 30 will never do so" [3]. Indeed, Einstein was merely 26 years old when he published his *Annus Mirabilis* papers. Yet, his observation about the link between youth and discovery was not merely autobiographical. Many of the physicists of his generation too made their defining discoveries very early in their career – Heisenberg and Dirac at 24; Pauli, Fermi, and Wigner at 25; Rutherford and Bohr at 28. But is youth a necessity for making an outstanding contribution to science? Clearly not. Alexander Fleming was 47 when he discovered penicillin. Luc Montagnier was 51 when he discovered HIV. And John Fenn was 67 when he first began to pursue the research that would later win him the Nobel Prize in chemistry. So, how is creativity, as captured by scientific breakthroughs, distributed across the lifespan of a career?

    The first part of this book will dive into these sets of fascinating questions regarding scientific careers. Indeed, as we survey our young and not so young colleagues doing groundbreaking work, we are prompted to ask: Are there quantitative patterns underlying when breakthrough work happens in a scientific career? What mechanisms drive the productivity and impact of a scientist? The chapters in this part will provide quantitative answers to these questions, offering insights that affect both the way we train scientists and the way we acknowledge and reward scientific excellence.

1 PRODUCTIVITY OF A SCIENTIST

Paul Erdős, arguably the most prolific mathematician in the twentieth century, was, by all accounts, rather eccentric. The Hungarian-born mathematician – who moved to the US before the start of WWII – lived out of a ragged suitcase that he famously dragged with him to scientific conferences, universities, and the homes of colleagues all over the world. He would show up unannounced on a colleague's doorstep, proclaim gleefully, "My mind is open." He then spent a few days working with his host, before moving on to surprise some other colleague at some other university. His meandering was so constant that it eventually earned him undue attention from the FBI. To his fellow mathematicians, he was an eccentric but lovable scientist. But to law enforcement officers during the Cold War, it was suspicious that he crossed the Iron Curtain with such ease. Indeed, Erdős was once arrested in 1941 for poking around a secret radio tower. "You see, I was thinking about mathematical theorems," he explained to the authorities in his thick Hungarian accent. It took decades of tracking for the Bureau to finally believe him, concluding that his rambling was indeed just for the sake of math.

His whole *life* was, too. He had no wife, no children, no job, not even a home to tie him down. He earned enough in guest lecturer stipends from universities and from various mathematics awards to fund his travels and basic needs. He meticulously avoided any commitment that might stand in the way of his work. Before he died in 1996 at the age of 83, Erdős had written or coauthored a stunning 1,475 academic papers in collaboration with 511 colleagues. If total publication counts as a measure of productivity, how does Erdős' number compare to the productivity of an ordinary scientist? It surely seems exceptional. But how exceptional?

1.1 How Much Do We Publish?

Scholarly publications are the primary mode of communication in science, helping disseminate knowledge. The productivity of a scientist captures the rate at which she adds units of knowledge to the field. Over the past century, the number of publications has grown exponentially. An important question is whether the growth in our body of knowledge is simply because there are now more scientists, or because each scientist produces more on average than their colleagues in the past.

An analysis of over 53 million authors and close to 90 million papers published across all branches of science shows that both the number of papers and scientists grew exponentially over the past century [4]. Yet, while the former grew slightly faster than the latter (Fig. 1.1a), meaning that the number of publications per capita has been decreasing over time, for each scientist, individual productivity has stayed quite stable over the past century. For example, the number of papers a scientist produces each year has hovered at around two for the entire twentieth century (Fig. 1.1b, blue curve), and has even increased slightly during the past 15 years. As of 2015, the typical scientist authors or coauthors about 2.5 papers per year. This growth in individual productivity has its origins in collaborations: Individual productivity is boosted as scientists end up on many more papers as coauthors (Fig. 1.1b, red curve). In other words, while in terms of how many scientists it takes to produce a paper, that number has been trending downwards over the past century, thanks to collaborative work individual productivity has increased during the past decade.

1.2 Productivity: Disciplinary Ambiguities

But, when it comes to a scientist's productivity, it's not easy to compare across disciplines. First, each publication may represent a unit of knowledge, but that unit comes in different sizes. A sociologist may

9 / Productivity of a Scientist

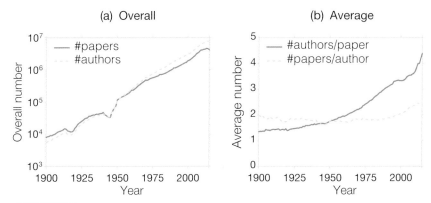

Figure 1.1 The growing number of scientists. (a) During the past century, both the number of scientists and the number of papers has increased at an exponential rate. (b) The number of papers coauthored by each scientist has been hovering around two during the past 100 years, and increased gradually in the past 15 years. This growth is a direct consequence of collaborative effects: Individual productivity is boosted as scientists end up on many more papers as coauthors. Similar trends were reported using data within a single field [5]. For physics, for example, the number of papers coauthored by each physicist has been less than one during the past 100 years, but increased sharply in the past 15 years. After Dong et al. [4] and Sinatra et al. [5].

not feel their theory is fully articulated unless the introduction of the paper spans a dozen pages. Meanwhile, a paper published in *Physical Review Letters*, one of the most respected physics journals, has a strict four-page limit, including figures, tables, and references. Also, when we talk about individual productivity, we tend to count publications in scientific journals. But in some branches of the social sciences and humanities, books are the primary form of scholarship. While each book is counted as one unit of publication, that unit is admittedly much more time-consuming to produce.

And then there is computer science (CS). As one of the youngest scientific disciplines (the first CS department was formed at Purdue University in 1962), computer science has adopted a rather unique publication tradition. Due to the rapidly developing nature of the field, computer scientists choose conference proceedings rather than journals as their primary venue to communicate their advances. This approach has served the discipline well, given everything that has been accomplished in the field – from the Internet to artificial intelligence – but it can be quite confusing to those outside the discipline.

Ignoring the varying publication conventions that characterize different disciplines can have serious consequences. For example, in 2017, the *US News and World Report* (US News), which develops authoritative ranking of colleges, graduate schools, and MBA programs around the world, published their first ranking of the world's best computer science departments. The ranking was so absurd that the Computing Research Association (CRA) had to put out a special announcement, calling it "nonsense" and "a grave disservice" to its readers.

How could an experienced organization specializing in ranking academic institutions get it so wrong? It turns out that *US News* calculated their rankings based on journal publications recorded by Web of Science, a procedure that served them well in all other disciplines. But, by ignoring peer-reviewed papers published in conferences, the *US News* rankings were completely divorced from computer scientists' own perceptions of quality and impact.

The productivity difference across disciplines can be quantified using data from the National Research Council on research doctorate programs in the US [6, 7]. Using the average number of publications by faculty in each department over a five-year period as a proxy, researchers find that the numbers ranged from 1.2 in history to 10.5 in chemistry. Even between similar disciplines we see large productivity differences. For example, within biological sciences, faculty productivity ranged from 5.1 in ecology to 9.5 in pharmacy.

Taken together, the data presented so far in this chapter make at least one message crystal clear: no matter how we measure it, the productivity of a typical scientist is nowhere near Erdős'. Indeed, his total – 1,475 papers – implies a staggering *two papers per month over a span of 60 years*. By contrast, a study focusing on more than 15 million scientists between 1996 and 2011, found that less than 1 percent of our colleagues managed to publish at least one paper every year [8]. Hence, only a small fraction of the scientific workforce can maintain a steady stream of publications. Interestingly, this small fraction contains the most high-impact researchers. Though they represent less than 1 percent of all publishing scientists, this stable core puts out 41.7 percent of all papers, and 87.1 percent of all papers with more than 1,000 citations. And if a productive scientist's pace lags, so does the impact of their contributions. Indeed, the average impact of papers published by a researcher is substantially lower if they skipped even a single year.

While Erdős is an outlier, his impressive productivity speaks to the enormous productivity differences among researchers. Why are there such differences? After all, we all have a 24-hour day to work with. So how can people like Erdős be so much more productive than their peers? To answer these questions, we need to visit the legendary Bell Laboratory in its heyday.

1.3 Productivity: The Difference

The career of William Shockley, the man who brought silicon to Silicon Valley, was not free of controversies. To be sure, his attempts to commercialize a new transistor design in the 1950s and 1960s transformed the Valley into the hotbed of electronics. Yet, his troubling advocacy for eugenics eventually isolated him from his colleagues, friends, and family. Shockley spent his most productive years at the Bell Laboratory, where he co-invented the transistor with John Bardeen and Walter Brattain. That discovery not only won the trio the 1956 Nobel Prize in Physics, it also began the digital revolution we continue to experience today.

While managing a research group at Bell Labs, Shockley became curious [9]: Were there measurable differences in the productivity of his fellow researchers? So he gathered statistics on the publication records of employees in national labs such as Los Alamos and Brookhaven. Once he charted the numbers, he was surprised by the outcome: The curve indicated that individual productivity, the number of papers published by a researcher, N, follows a lognormal distribution

$$P(N) = \frac{1}{N\sigma\sqrt{2\pi}} \exp\left(-\frac{(\ln N - \mu)^2}{2\sigma^2}\right). \tag{1.1}$$

Lognormal distributions are fat-tailed, capturing great variations in productivity. In other words, Shockley learned that most researchers publish very few papers, whereas a non-negligible fraction of scientists are orders of magnitude more productive than the average. Evidence for (1.1) is shown in Fig. 1.2, plotting the distribution of the number of papers written by all authors listed in INSPECT, together with a lognormal fit [10].

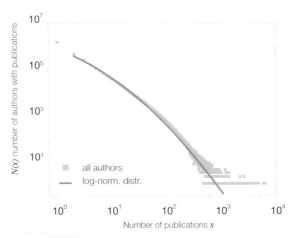

Figure 1.2 Productivity distribution. The blue symbols show the number of papers published by all authors listed in the INSPECT database of scientific and technical literature, in the period 1969–2004 (over 3 million authors). The red line corresponds to the lognormal fit to the data (1.1). After Fronczak et al. [10].

Box 1.1 The study of productivity has a long history [9–15]

In 1926, Alfred J. Lotka [11] observed that the number of papers produced by scientists follows a fat-tailed distribution. In other words, he found that a small fraction of scientists are responsible for the bulk of scientific literature. Lotka studied 6,891 authors listed in Chemical Abstracts publishing between 1907 and 1916, concluding that the number of authors making N contributions follows a power law

$$P(N) \sim N^{-\alpha}, \tag{1.2}$$

where the exponent $\alpha \approx 2$. A power law predicts that productivity has a long tail, capturing major variations among individuals. Note that it often requires a large amount of data to reliably distinguish a power law from a lognormal distribution [9], which Lotka did not have in 1926.

This lognormal distribution of productivity is rather odd, as Shockley quickly noticed. Indeed, in most competitive arenas, individual performance metrics almost always follow a narrow distribution. Think about running. At the Rio Olympics in 2016, Usain Bolt finished the 100-meter final in just 9.81 seconds. Justin Gatlin came in second and

Andre De Grasse in third, with running times 9.89 s and 9.91 s, respectively. These numbers are awfully close, reflecting a well-known fact that performance differences between individuals are typically bounded [16]. Similarly, Tiger Woods, even on his best day, only took down his closest contenders by a few strokes, and the fastest typist may only type a few words more per minute than a merely good one. The bounded nature of performance reminds us that it is difficult, if not impossible, to significantly outperform the competition in any domain. Yet, according to Fig. 1.2, this boundedness does not hold for scientific performance. Apparently, it *is* possible to be much better than your competitors when it comes to churning out papers. Why is that?

1.4 Why So Productive?

Shockley proposed a simple model to explain the lognormal productivity distribution he observed (Eq. 1.1) [9]. He suggested that in order to publish a paper, a scientist must juggle multiple factors, like:

F_1. Identify a good problem.
F_2. Make progress with it.
F_3. Recognize a worthwhile result.
F_4. Make a decision as to when to stop the research and start writing up the results.
F_5. Write adequately.
F_6. Profit constructively from criticism.
F_7. Show determination to submit the paper for publication.
F_8. Make changes if required by the journal or the referees.

If any of these steps fail, there will be no publication. Let us assume that the odds of a person clearing hurdle F_i from the list above is p_i. Then, the publication rate of a scientist is proportional to the odds of clearing each of the subsequent hurdles, that is $N \sim p_1 p_2 p_3 p_4 p_5 p_6 p_7 p_8$. If each of these odds are independent random variables, then the multiplicative nature of the process predicts that $P(N)$ follows a lognormal distribution of the form (1.1).

To understand where the outliers come from, imagine, that Scientist A has the same capabilities as Scientist B in all factors, except that A is twice as good at solving a problem (F_2), knowing when to stop (F_4), and determination (F_7). As a result, A's productivity will be eight times higher than B's. In other words, for each paper published by

Scientist B, Scientist A will publish eight. Hence small differences in scientists' ability to clear individual hurdles can together lead to large variations in overall productivity.

Shockley's model not only explains why productivity follows lognormal distribution, but it also offers a framework to improve our own productivity. Indeed, the model reminds us that publishing a paper does not hinge on a single factor, like having a great idea. Rather, it requires scientists to excel at multiple factors. When we see someone who is hyper-productive, we tend to attribute it to a single exceptional factor. Professor X is really good at coming up with new problems (F_1), or conveying her ideas in writing (F_5). The model suggests, however, that the outliers are unlikely to be explained by a single factor; rather, a researcher is most productive when she excels across many factors and fails in none.

The hurdle model indicates that a single weak point can choke an individual's productivity, even if he or she has many strengths. It also tells us that Erdős may have not been as super-human as we often think he was, or that his productivity might be attainable with careful honing of various skills. Indeed, if we could improve at every step of writing a paper, and even if it's just a tiny bit in each step, these improvements can combine to exponentially enhance productivity. Admittedly, this is easier said than done. But you can use this list to diagnose yourself: What step handicaps your productivity the most?

The remarkable variations in productivity have implications for reward. Indeed, Shockley made another key observation: while the productivity of a scientist is multiplicative, his salary – a form of reward often tied to performance – is additive. The highest paid employees earn at best about 50–100 percent more than their peers. There are many reasons why this is the case – it certainly seems fairer, and it helps ensure a collaborative environment. Yet, from a paper-per-dollar perspective, Shockley's findings raise some interesting questions about whether the discrepancy between additive salaries and multiplicative productivities could be exploited. Indeed, an institution may be better off employing a few star scientists, even if that means paying them a great deal more than their peers. Shockley's arguments are often used as a rationale for why top individuals at research-intensive institutions are offered much higher salaries and special perks, and why top departments within a university get disproportionately more funding and resources.

To be sure, gauging a career based on publication count alone grossly misrepresents how science works. Yet, individual productivity has been shown to closely correlate with the eminence of a scientist as well as her perceived contributions to the field. This pattern was documented by Wayne Dennis, dating back at least to 1954 [1], when he studied 71 members of the US National Academy of Sciences and eminent European scientists. He found that, almost without exception, highly productive individuals have also achieved scientific eminence, as demonstrated by their listing in the *Encyclopedia Britannica* or in histories of important developments they have contributed to the sciences. Higher productivity has been shown to increase the odds of receiving tenure [17], and of securing funding for future research [18]. At the institutional level, the publication rates of the faculty are not only a reliable predictor of a program's reputation, they also influence the placement of graduates into faculty jobs [19].

In sum, sustained high productivity is rare, but it correlates with scientific impact and eminence. Given this evidence, it may appear that productivity is the key indicator for a meaningful career in science. Yet, as we show in the following chapters, among the many metrics used to quantify scientific excellence, productivity is the least predictive. The reason is simple: While great scientists tend to be very productive, not all scientists who are productive make long-lasting contributions. In fact, most of them do not. Multiple paths can lead to achieving high productivity. For example, lab technicians in certain fields may find their names on more than a hundred – or sometimes as many as a thousand – papers. Hence, they appear to be exceptionally prolific based on their publication counts, but are rarely credited as the intellectual owner of the research. The way people publish is also changing [20]. Coauthorship is on the rise, as are multiple publications on the same data. There have also been more discussions about LPUs, which stands for least publishable unit [20] or the "salami publishing" approach, which could further contribute to inflated productivity counts.

So, if productivity is not the defining factor of a successful career, what is?

Box 1.2 Name disambiguation

Our ability to accurately track individual productivity relies on our skill to identify the individual(s) who wrote a paper and all other work that belongs to that individual [21, 22]. This seemingly simple task represents a major unsolved problem [21–23], limited by four challenges. First, a single individual may appear in print under multiple names because of orthographic and spelling variants, misspellings, name changes due to marriage, religious conversion, gender reassignment, or the use of pen names. Second, some common names can be shared by multiple individuals. Third, the necessary metadata is often incomplete or missing. This includes cases where publishers and bibliographic databases failed to record authors' first names, their geographical locations, or other identifying information. Fourth, an increasing percentage of papers is not only multi-authored, but also represents multidisciplinary and multi-institutional efforts. In such cases, disambiguating some of the authors does not necessarily help assign the remaining authors.

While multiple efforts are underway to solve the name disambiguation problem, we need to be somewhat mindful about the results presented in this and following chapters, as some conclusions may be affected by the limitations in disambiguation. In general, it is easier to disambiguate productive scientists, who have a long track record of papers, compared with those who have authored only a few publications. Therefore, many studies focus on highly productive scientists with unusually long careers instead of "normal" scientists.

2 THE *h*-INDEX

Lev Landau, a giant of Russian physics, kept a handwritten list in his notebook, ranking physicists on a logarithmic scale of achievement and grading them into "leagues" [24]. According to Landau, Isaac Newton and Albert Einstein belonged to the highest rank, above anyone else: he gave Newton the rank 0 and Einstein a 0.5. The first league, a rank of 1, contains the founding fathers of quantum mechanics, scientists like Niels Bohr, Werner Heisenberg, Paul Dirac, and Erwin Schrödinger. Landau originally gave himself a modest 2.5, which he eventually elevated to 2 after discovering superfluidity, an achievement for which he was awarded the Nobel Prize. Landau's classification system wasn't limited to famous scientists, but included everyday physicists, who are given a rank of 5. In his 1988 talk "My Life with Landau: Homage of a 4 1/2 to a 2," David Mermin, who coauthored the legendary textbook *Solid State Physics*, rated himself a "struggling 4.5" [25].

When scientists leave league 5 behind and start approaching the likes of Landau and other founders of a discipline, it's obvious that their research has impact and relevance. Yet for the rest of us, things are somewhat blurry. How do we quantify the cumulative impact of an individual's research output? The challenge we face in answering this question is rooted in the fact that an individual's scientific performance is not just about how many papers one publishes, but a convolution of productivity and impact, requiring us to balance the two aspects in a judicious manner.

Of the many metrics developed to evaluate and compare scientists, one stands out in its frequency of use: the *h*-index, proposed by

Jorge E. Hirsch in 2005 [26]. What is the h-index, and how to calculate it? Why is it so effective in gauging scientific careers? Does it predict the future productivity and impact of a scientist? What are its limitations? And how do we overcome these limitations? Answering these questions is the aim of this chapter.

2.1 The *h*-Index: Definitions and Implications

The index of a scientist is h if h of her papers have at least h citations and each of the remaining papers have less than h citations [26]. For example, if a scientist has an h-index of 20 (h = 20), it means that she has 20 papers with more than 20 citations, and the rest of her papers all have less than 20 citations. To measure h, we sort an individual's publications based on her citations, going from the most cited paper to the least cited ones. We can plot them on a figure, that shows the number of citations of each paper, resulting in a monotonically decreasing curve. Fig. 2.1 uses the careers of Albert Einstein and Peter Higgs as case studies showing how to calculate their h-index.

Is an h-index of 8, for example, impressive or modest? What is the expected h-index of a scientist? To answer these questions, let's take a look at a simple but insightful model proposed by Hirsch [26]. Imagine that a researcher publishes n papers each year. Let us also assume that each paper earns c new citations every year. Hence a paper's citations increase linearly with its age. This simple model predicts the scientist's time dependent h-index as

$$h = \frac{c}{1 + c/n} t. \tag{2.1}$$

Therefore, if we define

$$m \equiv \frac{1}{1/c + 1/n}, \tag{2.2}$$

we can rewrite (2.1) as

$$h = mt, \tag{2.3}$$

indicating that a scientist's h-index increases approximately linearly with time. Obviously, researchers don't publish exactly the same number of papers every year (see Chapter 1), and citations to a paper follow varied temporal trajectories (as we will cover in

19 / The h-Index

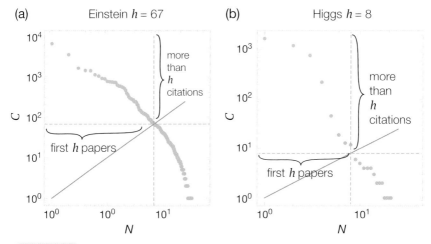

Figure 2.1 The h-index of Albert Einstein (a) and Peter Higgs (b). To calculate the h-index, we plot the number of citations versus paper number, with papers listed in order of decreasing citations. The intersection of the 45° line with the curve gives h. The total number of citations is the area under the curve [26]. According to Microsoft Academic Graph, Einstein has an h-index of 67, and Higgs 8. The top three most cited papers by Einstein are: (1) Can quantum mechanical description of physical reality be considered complete, *Physical Review*, 1935; (2) Investigations on the theory of Brownian movement, *Annalen der Physik*, 1905; and (3) On the electrodynamics of moving bodies, *Annalen der Physik*, 1905. The top three for Higgs are: (1) Broken symmetries and the masses of gauge bosons, *Physical Review Letters*, 1964; (2) Broken symmetries, massless particles and gauge fields, *Physics Letters*, 1964; (3) Spontaneous symmetry breakdown without massless bosons, *Physical Review*, 1966.

Chapter 19). Yet, despite the model's simplicity, the linear relationship predicted by (2.3) holds up generally well for scientists with long scientific careers [26].

This linear relationship (2.3) has two important implications:

(1) If a scientist's h-index increases roughly linearly with time, then its speed of growth is an important indicator of her eminence. In other words, the differences between individuals can be characterized by the slope, m. As (2.2) shows, m is a function of both n and c. So, if a scientist has higher productivity (a larger n), or if her papers collect more citations (higher c), she has a higher m. And the higher the m, the more eminent is the scientist.

(2) Based on typical values of m, the linear relationship (2.3) also offers a guideline for how a typical career should evolve. For

example, Hirsch suggested in 2005 that for a physicist at major research universities, $h \approx 12$ might be a typical value for achieving tenure (i.e., the advancement to associate professor) and that $h \approx 18$ might put a faculty member into consideration for a full professorship. Fellowship in the American Physical Society might typically occur around $h \approx 15$–20, and membership in the US National Academy of Sciences may require $h \approx 45$ or higher.

Since its introduction, the h-index has catalyzed a profusion of metrics and greatly popularized the idea of using objective indicators to quantify nebulous notions of scientific quality, impact or prestige [27]. As a testament to its impact, Hirsh's paper, published in 2005, had been cited more than 8,000 times as of the beginning of 2019, according to Google Scholar. It even prompted behavioral changes – some ethically questionable – with scientists adding self-citations for papers on the edge of their h-index, in hopes of boosting it [28–30]. Given its prevalence, we must ask: can the h-index predict the future impact of a career?

Box 2.1 The Eddington number

The h-index for scientists is analogous to the Eddington number for cyclists, named after Sir Arthur Eddington (1882–1944), an English astronomer, physicist, and mathematician, famous for his work on the theory of relativity. As a cycling enthusiast, Eddington devised a measure of a cyclist's long-distance riding achievements. The Eddington number, E, is the number of days in your life when you have cycled more than E miles. Hence an Eddington number of 70 would mean that the person in question has cycled at least 70 miles a day on 70 occasions. Achieving a high Eddington number is difficult, since jumping from, say, 70 to 75 may require more than 5 new long-distance rides. That's because any rides shorter than 75 miles will no longer be included. Those hoping to increase their Eddington number are forced to plan ahead. It might be easy to achieve an E of 15 by doing 15 trips of 15 miles – but turning that $E = 15$ into an $E = 16$ could force a cyclist to start over, since an E number of 16 only counts trips of 16 miles or more. Arthur Eddington, who reached an $E = 87$ by the time he died in 1944, clearly understood that if he wanted to achieve a high E number, he had to start banking long rides early on.

2.2 The Predictive Power of the *h*-Index

To understand the value of the *h*-index, let's take a look at the "usual suspects" – metrics that are commonly used to evaluate a scientist's performance, and review their strengths and limitations [26].

(1) Total number of publications (N).
   Advantage: Measures the productivity of an individual.
   Disadvantage: Ignores the impact of papers.
(2) Total number of citations (C).
   Advantage: Measures a scientist's total impact.
   Disadvantage: It can be affected by a small number of big hits, which may not be representative of the individual's overall career, especially when these big hits were coauthored with others. It also gives undue weight to highly cited reviews as opposed to original research contributions.
(3) Citations per paper (C/N).
   Advantage: Allows us to compare scientists of different ages.
   Disadvantage: Outcomes can be skewed by highly cited papers.
(4) The number of "significant papers," with more than c citations.
   Advantage: Eliminates the disadvantages of (1), (2), (3), and measure broad and sustained impact.
   Disadvantage: The definition of "significant" introduces an arbitrary parameter, which favors some scientists or disfavors others.
(5) The number of citations acquired by each of the q most-cited papers (for example, $q = 5$).
   Advantage: Overcomes many of the disadvantages discussed above.
   Disadvantage: Does not provide a single number to characterize a given career, making it more difficult to compare scientists to each other. Further, the choice of q is arbitrary, favoring some scientists while handicapping others.

The key advantage of the *h*-index is that it *sidesteps all of the disadvantages* of the metrics listed above. But, is it more effective at gauging the impact of an individual's work? When it comes to evaluating the predictive power of metrics, two questions are often the most relevant.

Q1: Given the value of a metric at a certain time t_1, how well does it predict the value of itself or of another metric at a future time t_2?

This question is especially interesting for hiring decisions. For example, if one consideration regarding a faculty hire is the likelihood of the candidate to become a member of the National Academy of Sciences 20 years down the line, then it would be useful to rank the candidates by their projected *cumulative* achievement after 20 years. Hirsch tested Q1 by selecting a sample of condensed matter physicists and looked at their publication records during the first 12 years of their career and in the subsequent 12 years [31]. More specifically, he calculated four different metrics for each individual based on their career records in the first 12 years, including the *h*-index (Fig. 2.2a), the total number of citations

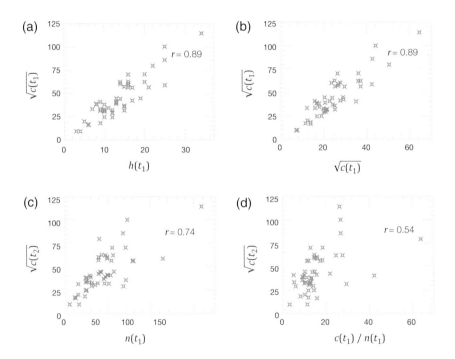

Figure 2.2 Quantifying predictive power of the *h*-index. Scatter plots compare the total number of citations, C, after $t_2 = 24$ years vs. the value of the various indicators at $t_1 = 12$ year for each individual within the sample. Hirsch hypothesized C may grow quadratically with time, and hence used its square root when calculating the total number of citations. By calculating the correlation coefficient, he found that the *h*-index (a) and the number of citations at t_1 (b) are the best predictors of the future cumulative citations at t_2. The number of papers correlates less (c), and the number of citations per paper performs the worst (d). After Hirsch [31].

(Fig. 2.2b), the total number of publications (Fig. 2.2c), and the average number of citations per paper (Fig. 2.2d). He then asked if we want to select candidates that have the most total citations by year 24, which one of the four indicators gives us the best chance? By measuring the correlation coefficient between future cumulative citations at time t_2 and four different metrics calculated at time t_1, he found that the h-index and the number of citations at time t_1 turn out to be the best predictors (Fig. 2.2).

While Fig. 2.2 shows that the h-index predicts cumulative impact, in many cases it's the future scientific output that matters the most. For example, if we're deciding who should get a grant, how many more citations an applicant's *earlier* papers are expected to collect in the next few years is largely irrelevant. We're concerned, instead, with papers that the potential grantee has not yet written and the impact of those papers. Which brings us to Q2:

Q2: How well do the different metrics predict *future* scientific output?

To answer Q2, we need to use indicators obtained at t_1 to predict scientific achievement occurring only in the subsequent period, thereby omitting all citations to work performed prior to t_1. Hirsch repeated the similar prediction task for the four metrics, but this time used each of them to predict total citations accrued by papers published only in the next 12 years. Naturally, this is a more difficult task, but an important one for allocating research resources. Hirsch found that the h-index again emerges as the best predictor for achievement incurred purely in future time frame [31].

These findings indicate that two individuals with similar h are comparable in terms of their overall scientific achievement, even if their total number of papers or citations are quite different. Conversely, two individuals of the same scientific age can have a similar number of total papers or citation counts but very different h values. In this case, the researcher with the higher h is typically viewed by the community as the more accomplished. Together, these results highlight the key strength of the h-index: When evaluating scientists, it gives an easy but relatively accurate estimate of an individual's overall scientific achievements. Yet at the same time, we must also ask: What are the limitations of the h-index?

> **Box 2.2 The birth of the *h*-index**
>
> Since its inception, the *h*-index has been an integral part of a scientific life. Its exceptional influence prompted us to reach out to Jorge Hirsch to ask how he arrived to the measure. He kindly responded, writing:
>
> > I thought about it first in mid 2003, over the next weeks I computed the *h*-index of everybody I knew and found that it usually agreed with the impression I had of the scientist. Shared it with colleagues in my department, several found it interesting.
> >
> > Mid June 2005 I wrote up a short draft paper, sent it to 4 colleagues here. One skimmed over it, liked it and made some suggestions, one liked some of it and was nonplussed by some of it, two didn't respond. So I wasn't sure what to do with it.
> >
> > Mid July 2005 I got out of the blue an email from Manuel Cardona in Stuttgart saying he had heard about the index from Dick Zallen at Virginia Tech who had heard about it from one of my colleagues at UCSD (didn't say who but I can guess). At that point I decided to clean up the draft and post it in arXiv, which I did August 3, 2005, still was not sure what to do with it. Quickly got a lot of positive (and some negative) feedback, sent it to *PNAS* August 15.

2.3 Limitations of the *h*-Index

The main street of College Hill in Easton, Pennsylvania – the home of the Lafayette College – is named after James McKeen Cattell. As an American psychologist, Cattell played an instrumental role in establishing psychology as a legitimate science, advocacy that prompted the *New York Times* to call him "the dean of American science" in his obituary.

While many have thought of developing new metrics to systemically evaluate their fellow researchers, Cattell was the first to popularize the idea of ranking scientists. He wrote in his 1910 book, *American Men of Science: A Biographical Directory* [32]: "It is surely time for scientific men to apply scientific method to determine the circumstances that promote or hinder the advancement of science." So, today's obsession of measuring impact using increasingly sophisticated yardsticks is by no means a modern phenomenon. Scientists have been sizing up their colleagues since the beginning of the discipline itself. A century after Cattell's book, the need and the rationale for a reliable toolset to evaluate scientists has not changed [33].

As the h-index has become a frequently used metric of scientific achievements, we must be mindful about its limitations. For example, although a high h is a somewhat reliable indicator of high accomplishment, the converse is not necessarily always true [31]: an author with a relatively low h can achieve an exceptional scientific impact with a few seminal papers, such as the case of Peter Higgs (Fig. 2.1b). Conversely, a scientist with a high h achieved mostly through papers with many coauthors would be treated overly kindly by his or her h. Furthermore, there is considerable variation in citation distributions even within a given subfield, and subfields where large collaborations are typical (e.g., high-energy experimental physics) will exhibit larger h values, suggesting that one should think about how to normalize h to more effectively compare and evaluate different scientists.

Next we discuss a few frequently mentioned limitations of the h-index, along with variants that can – at least to a certain degree – remedy them.

- **Highly cited papers.** The main advantage of the h-index is that its value is not boosted by a single runaway success. Yet this also means that it neglects the most impactful work of a researcher. Indeed, once a paper's citations get above h, its relative importance becomes invisible to the h-index. And herein lies the problem – not only do outlier papers frequently define careers, they arguably are what define science itself. Many remedies have been proposed to correct for this [34–39], including the g-index (the highest number g of papers that *together* received g^2 or more citations [40, 41]) and the o-index (the geometric mean of the number of citations gleaned by a scientist's highest cited papers c^* and her h-index: $o = \sqrt{c^* h}$ [42]). Other measures proposed to correct this bias include a-index [36, 38]; $h(2)$-index [39]; h_g-index [34]; q^2-index [37]; and more [35].
- **Inter-field differences.** Molecular biologists tend to get cited more often than physicists who, in turn, are cited more often than mathematicians. Hence biologists typically have higher h-index than physicists, and physicists tend to have an h-index that is higher than mathematicians. To compare scientists across different fields, we

must account for the field-dependent nature of citations [43]. This can be achieved by the h_g-index, which rescales the rank of each paper n by the average number of papers written by author in the same year and discipline, n_0 [43] or the h_s-index, which normalizes the h-index by the average h of the authors in the same discipline [44].

- **Time dependence.** As we discussed in Chapter 2.2, the h-index is time dependent. When comparing scientists in different career stages, one can use the m quotient (2.2) [26], or contemporary h-index [45].
- **Collaboration effects.** Perhaps the greatest shortcoming of the h-index is its inability to discriminate between authors that have very different coauthorship patterns [46–48]. Consider two scientists with similar h indices. The first one is usually the intellectual leader of his/her papers, mostly coauthored with junior researchers, whereas the second one is mostly a junior author on papers coauthored with eminent scientists. Or consider the case where one author always publishes alone whereas the other one routinely publishes with a large number of coauthors. As far as the h-index is concerned, all these scientists are indistinguishable. Several attempts have been proposed to account for the collaboration effect, including fractionally allocating credit in multi-authored papers [48–50], and counting different roles played by each coauthor [51–54] by for example differentiating the first and last authorships. Hirsch himself has also repeatedly acknowledged this issue [46, 47], and proposed the h_α-index to quantify an individual's scientific leadership for their collaborative outcomes [47]. Among all the papers that contribute to the h-index of a scientist, only those where he or she was the most senior author (the highest h-index among all the coauthors) are counted toward the h_α-index. This suggests that a high h-index in conjunction with a high h_α/h ratio is a hallmark of scientific leadership [47].

In addition to these variations of the h-index, there are other metrics to quantify the overall achievement of individual scientists, including the i10-index, used exclusively by Google Scholar [55], which computes the number of articles with at least 10 citations each; or the SARA method [56], which uses a diffusion algorithm that mimics the spreading of scientific credits on the citation network to quantify an individual's scientific eminence. Despite the multitude of metrics attempting to correct the shortcomings of the h-index, to date no other bibliometric index has emerged as preferable to the h-index, cementing the status of the h-index as a widely used indicator of scientific achievement.

As we dug deeper into h-index and the voluminous body of work motivated by it, it was easy to forget a perhaps more important point: No scientist's career can be summarized by a single number. Any metric, no matter how good it is at achieving its stated goal, has limitations that must be recognized before it is used to draw conclusions about a person's productivity, the quality of her research, or her scientific impact. More importantly, a scientific career is not just about discoveries and citations. Rather, scientists are involved in much broader sets of activities including teaching, mentoring, organizing scientific meetings, reviewing, and serving on editorial boards, to name a few. As we encounter more metrics for scientific eminence, it's important to keep in mind that, while they may help us understand certain aspects of scientific output, none of them alone can capture the diverse contributions scientists make to our community and society [57, 58]. Just as Einstein cautioned: "Many of the things you can count, don't count. Many of the things you can't count, do count."

Therefore, we must keep in mind that the h-index is merely a proxy to quantify scientific eminence and achievement. But the problem is, in science, status truly matters, influencing the perception of quality and importance of one's work. That's what we will focus on in the next chapter, asking if and when status matters, and by how much.

3 THE MATTHEW EFFECT

Lord Rayleigh is a giant of physics, with several laws of nature carrying his name. He is also known beyond the profession thanks to Rayleigh scattering, which answers the proverbial question, "Why is the sky blue?" Rayleigh was already a respected scientist when, in 1886, he submitted a new paper to the *British Association for the Advancement of Science* to discuss some paradoxes of electrodynamics. The paper was promptly rejected on the grounds that it did not meet the journal's expectation of relevance and quality. Yet, shortly after the decision, the editors reversed course. Not because anything changed about the paper itself. Rather, it turns out that Rayleigh's name had been inadvertently omitted from the paper when it was first submitted. Once the editors realized it was Rayleigh's work, it was immediately accepted with profuse apologies [59, 60]. In other words, what was initially viewed as the scribblings of some "paradoxer," suddenly became worth publishing once it became clear that it was the work of a world-renowned scientist.

This anecdote highlights a signaling mechanism critical in science: the role of scientific reputation. Robert K. Merton in 1968 [60] called this the Matthew effect after a verse in the biblical Gospel of Matthew pertaining to Jesus' parable of the talents: "For to everyone who has will more be given, and he will have an abundance. But from the one who has not, even what he has will be taken away." The Matthew effect as a concept has been independently discovered in multiple disciplines over the last century, and we will encounter it again in Chapter 17, when we discuss citations. In the context of careers, the

Matthew effect implies that a scientist's status and reputation alone can bring additional attention and recognition. This means that status not only influences the community's perception of the scientist's credibility, playing an important role in how her work is evaluated, but it also translates into tangible assets – from research funding to access to outstanding students and collaborators – which in turn further improve her reputation. The goal of this chapter is to unpack the role of the Matthew effect in careers. When does it matter? And to what extent?

3.1 What's in a Name?

The Internet Engineering Task Force (IETF) is a community of engineers and computer scientists who develop the protocols that run the Internet. To ensure quality and functionality, engineers must submit all new protocols as manuscripts that undergo rigorous peer review. For a while, each manuscript included the name of every author. However, beginning in 1999, some manuscripts replaced the full author list with a generic "et al.," concealing the name of some authors from the review committee.

By comparing cases where well-known authors were hidden by the et al. label with those where the hidden names were little-known, researchers effectively conducted a real-world Lord Rayleigh experiment [61]. They found that when an eminent name was present on a submission, like the chair of a working group, which signals professional standing, the submission was 9.4 percent more likely to be published. However, the "chair effect" declined by 7.2 percent when the senior author's name was masked by the et al. label. In other words, name-based signaling accounts for roughly 77 percent of the benefits of having an experienced author as a coauthor on the manuscript.

Interestingly, when the analysis was restricted to a small pool of manuscripts that were "pre-screened," or closely scrutinized, the author name premium disappeared. This suggests that the status effect only existed when the referees were dealing with high submission rates. In other words, when the reviewers do actually read the manuscript, carefully judging their content, status signals tend to disappear.

Given the exponential growth of science, we frequently encounter the "too many to read" situations. Yet, typically, peer review is a rather involved process, with multiple rounds of communication between authors and expert reviewers, suggesting that the status signaling may be less of a concern for scientific manuscripts. Indeed, through those rebuttals and revisions, an objective assessment of the work is expected to prevail. Yet, as we see next, the status effect is rarely eliminated.

Whether an author's status affects the *perceived* quality of his/her papers has been long debated in the scientific community. To truly assess the role of status, we need randomized control experiments, where the same manuscript undergoes two separate reviews, one in which the author identities are revealed and another in which they are hidden. For obvious ethical and logistical reasons, such an experiment is difficult to carry out. Yet, in 2017, a team of researchers at Google were asked to co-chair the program of the Tenth Association for Computing Machinery International Conference on Web Search and Data Mining (WSDM), a highly selective computer science conference with a 15.6 percent acceptance rate. The researchers decided to use the assignment as a chance to assess the importance of status for a paper's acceptance [62].

There are multiple ways to conduct peer review. The most common is the "single-blind" review, when the reviewers are fully aware of the identity of the authors and the institution where they work, but, the authors of the paper are not privy to the reviewer's identity. In contrast, in "double-blind" review, neither the authors nor the reviewers know each other's identity. For the 2017 WSDM conference the reviewers on the program committee were randomly split into a single-blind and a double-blind group. Each paper was assigned to four reviewers, two from the single-blind group and two from the double-blind group. In other words, two groups of referees were asked to independently judge the same paper, where one group was aware of who the authors were, while the other was not.

Given the Lord Rayleigh example, the results were not surprising: Well-known author – defined as having at least three papers accepted by previous WSDM conferences and at least 100 computer science papers in total – were 63 percent more likely to have the paper accepted under single-blind review than in double-blind review. The papers under review in these two processes were exactly the same, therefore, the difference in acceptance rate can only be explained by author identity. Similarly, authors from top universities had a 58 percent increase in acceptance once their affiliation was known. Further, for authors working at Google, Facebook, or Microsoft, considered prestigious institutions in computer science, the acceptance rate more than doubled, increasing by 110 percent.

These results indicate that in science we *do* rely on reputation – captured by both author identity and institution prestige – when we judge a scientific paper. While we may debate whether the use of such information helps us make better or worse decisions, status signaling plays a key role in getting published in prestigious venues.

Box 3.1 Double blind vs. single blind

If two papers are identical except that one is written by an unknown scientist while the other by a researcher of considerable reputation, they obviously should have an equal chance of being published. Yet, this may only happen if the reviews are double-blind. Isn't this a strong reason for all science to switch to a double-blind review? New evidence shows that this simple question may not have a simple answer.

An experiment performed by *Nature* shows that making double-blind review optional does not solve the problem [63]. *Nature* started to offer the option of double-blind review in 2015. Analyzing all papers received between March 2015 and February 2017 by 25 *Nature*-branded journals, researchers found that corresponding authors from less prestigious institutions were more likely to choose double-blind review, presumably as an attempt to correct for pre-existing biases. However, the success rate at both first decision and peer review was significantly lower for double-blind than single-blind reviewed papers.

Take *Nature*, the flagship journal, as an example. Under a single-blind process, the odds of your paper to be sent out for review is 23 percent. But if you opt for a double-blind review, those odds drop to 8 percent. The prospects don't improve once the paper is reviewed: The likelihood that your paper is accepted after review is around 44 percent for single-blind submissions but only 25 percent for double-blind papers. In other words, referees are more critical if they are unaware of an author's identity. And these differences matter: If you multiply the probabilities of being sent out for review and acceptance after review, the chances of your paper being accepted in *Nature* through the double-blind review process is a staggering 2 percent. While a single-blind review is still a long shot, it has a success rate of 10.1 percent. In other words, your odds of acceptance may drop as much as five-fold by simply selecting the double-blind option.

> A possible explanation for the observed difference is the quality of papers [63]. Indeed, papers by less-known authors or from less prestigious institutions may not report research of the same quality as those coming from experienced authors from elite institutions. This argument is refuted by a randomized experiment conducted at the *American Economic Review (AER)* [64], an influential journal in economics. From May 1987 to May 1989, *AER* ran a randomized experiment on submitted papers, assigning one-half to a double-blind review process and the other half to single-blind one. Because the experiment was randomized, there was no quality difference between the two piles. Yet, the acceptance rate was still much lower for papers in the double-blind group.
>
> These results suggest positive discrimination as a more likely explanation: authors with perceived status are subject of the benefit of the doubt, whereas those without the status get extra scrutiny on issues from research design to methodology. Yet, it is difficult to find objective reasons why double-blind review should not be the norm of scientific publishing. Its widespread use would level the playing field, lowering the hold of researchers with status – whether earned through the impact of previous work, or simply through affiliation with highly prestigious institutions.

3.2 Boosting Impact

A scientist's reputation facilitates the publication of her paper, but does it also affect the long-term impact of their discoveries? While acceptance only requires a momentary consensus among reviewers, its impact is driven by the broader consensus of the scientific community, who may choose to build further on the work, or may simply ignore it. Do papers published by a well-known scientist also enjoy an impact premium?

But, how do we measure reputation? A good starting point may be $c_i(t)$, representing the total citation count of all papers an author has published prior to a new paper's publication [65, 66]. Indeed, $c_i(t)$ combines both productivity (the number of papers published by an author) and impact (how often these papers are cited by others), which together offer a reasonable proxy of the author's name recognition within her research community.

The rate at which a paper acquires new citations tends to be proportional to how many citations the paper has already collected [67, 68]. Indeed, highly cited papers are more read, hence are more likely to be cited again. This phenomenon is called preferential attachment, which we will discuss again in detail in Chapter 17. To see how an author's reputation affects the impact of her publications, we can measure the early citation premium for well-known authors [65]. For example, for a group of well-known physicists, their paper has acquired around 40 citations ($c_x \approx 40$) before preferential attachment turns on (Fig. 3.1). In contrast, for junior faculty in physics (assistant professors), c_x drops from 40 to 10. In other words, right after its publication, a senior author's paper appears four times more likely to be cited than a junior author's.

Figure 3.1 suggests that reputation plays an important role early on, when the number of citations is small (i.e., when $c < c_x$). Yet, with time, the reputation effect fades away, and the paper's long-term impact is primarily driven by mechanisms inherent to *papers* rather than their *authors*. In other words, well-known authors enjoy an early citation premium, representing better odds of their work to be noticed by the community. This leads to a leg-up in early citations. But with time, this reputation effect vanishes, and preferential attachment takes over, whose rate is driven primarily by the collective perception of the inherent value of the discovery.

The reputation boost discussed above is not limited to new papers. Eminence can spill over to earlier works as well, boosting their impact. Sudden recognitions, like receiving the Nobel Prize, allow us to quantify this effect. Consider, for example, John Fenn, who received the 2002 Nobel Prize in chemistry for the development of the

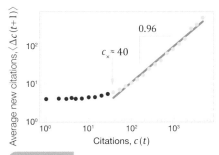

Figure 3.1 The cross-over effect of reputation on citations. The linear attachment rate breaks down for $c < c_x$, suggesting that additional forces provide a citation boost which elevates $c(t)$ to deviate from what is predicted by the pure preferential attachment mechanism. Datasets include 100 top-cited physicists, and another 100 highly prolific physicists. After Petersen et al. [65].

electrospray ionization technique. His original discovery, published in *Science* in 1989 [69], is Fenn's most cited work, collecting close to 8,500 citations by 2018 according to Google Scholar. But as his landmark paper started to collect citations at an exceptional rate following its publication, the citation rates of several of Fenn's older papers also started to grow at a higher pace. Analyses of 124 Nobel laureates show that this boost is common [70]: The publication of a major discovery increases the citation rates of papers the author published *before*. Interestingly, the older papers that enjoyed the citation boosts are *not necessarily related* to the topic of the new discovery. In other words, reputational signaling operates by bringing professional attention to the individual. Consequently, when an author becomes prominent in one area of science, her reputation may be extended to her other line of work, even in unrelated fields.

Box 3.2 From boom to bust: The reverse Matthew effect

If a major breakthrough blesses both past and future scholarship, what does a scandal do to a career? Scientists are certainly fallible, and the scientific community regularly confronts major mistakes or misconduct. These incidents lead to retractions of articles, particularly in top journals [71], where they receive enhanced scrutiny. To what degree does a retracted paper affect a scientific career? Are eminent authors affected more or less severely than their junior colleagues? While retractions are good for science, helping other researchers avoid false hypotheses, retractions are never good for the authors of the retracted paper: they experience a spillover, leading to citation losses to their prior body of work as well [72–74]. The negative impact is not distributed equally, however: Eminent scientists are more harshly penalized for their retracted papers than when retractions happen to their less-distinguished peers [74]. Importantly, this conclusion only holds when the retractions involve fraud or misconduct. In other words, when the retraction is perceived to be the consequence of an "honest mistake," the penalty differential between high- and low-status authors disappears [74].

When a senior and junior scientists are on the same retracted paper, however, the status penalty becomes quite different [75]: Senior authors often escape mostly unscathed, whereas their junior collaborators carry the blame, sometimes even to a career-ending degree. We will return to this effect in Chapter 13, where we explore the benefits and the drawbacks of collaborative work.

3.3 Is it Really the Matthew Effect After All?

Great scientists are seldom one-hit wonders [60, 76]. Newton is a prime example: beyond the Newtonian mechanics, he developed the theory of gravitation, calculus, laws of motion, optics, and optimization. In fact, well-known scientists are often involved in multiple discoveries, another phenomenon potentially explained by the Matthew effect. Indeed, an initial success may offer a scientist legitimacy, improve peer perception, provide knowledge of how to score and win, enhance social status, and attract resources and quality collaborators, each of these payoffs further increasing her odds of scoring another win. Yet, there is an appealing alternative explanation: Great scientists have multiple hits and consistently succeed in their scientific endeavors simply because they're exceptionally talented. Therefore, future success again goes to those who have had success earlier, *not* because of advantages offered by the previous success, but because the earlier success was indicative of a hidden talent. The Matthew effect posits that success *alone* increases the future probability of success, raising the question: Does status dictate outcomes, or does it simply reflect an underlying talent or quality? In other words, is there really a Matthew effect after all?

Why should we care about which is the more likely explanation, if the outcome is the same? Indeed, independent of the mechanism, people who have previously succeeded are more likely to succeed again in the future. But, if innate differences in talent is the only reason why some people succeed while others don't, it means that the deck is simply stacked in favor of some – at the expense of others – from the outset. If, however, the Matthew effect is real, each success you experience will better your future chances. You may not be Einstein, but if you are lucky to get that early win, you may narrow the gap between yourself and someone of his eminence, as your success snowballs.

Unfortunately, it is rather difficult to distinguish these two competing theories, as they yield similar empirical observations. One test of these contrasting hypotheses was inspired by the French Academy's mythical "41st chair." The Academy decided early on to have only 40 seats, limiting its membership to 40 so-called "immortals," and would only consider nominations or applications for new members if one of the seats became vacant through the death of a member. Given this restriction, many deserving individuals were never elected into the Academy, being eternally delegated to the 41st chair. It's a crowded seat, shared by true immortals like Descartes, Pascal, Molière, Rousseau, Saint-Simon, Diderot, Stendahl, Flauberta, Zola, and Proust [60]. At the same time, many of those who did hold a seat in the esteemed club are (unfortunately) utterly irrelevant to us today. With time, the 41st chair became a symbol of the many talented scientists who *should* have been, but were never, recognized as giants of their discipline.

But, does it actually matter if someone is formally recognized or not? Indeed, how does the post-award perception of major prizewinners compare to scientists who had comparable performance, but who were not officially recognized? In other words, how does the career of those that occupied the 41st chair differed, had they been elected to the French Academy? The answer is provided by a study, exploring the impact of a major status-conferring prize [77].

As a prestigious private funding organization for biomedical research in the United States, the Howard Hughes Medical Institute (HHMI) selects "people, not projects," generously supporting scientists rather than awarding them grants for specific peer-reviewed research proposals. The HHMI offers about US$1 million per investigator each year, providing long-term, flexible funding that allows awardees the freedom to follow their instincts, and if necessary, change research directions. Beyond the monetary freedom, being appointed an HHMI investigator is seen as a highly prestigious honor. To measure the impact of the HHMI award, the challenge is to create a control group of scientists who were close contenders but who were not selected for the award and compare their scientific outputs with those of the HHMI investigators.

But, let's assume that we identify this control group of scientists, and do find evidence that HHMI investigators have more impact. How can we know that the difference is purely because of their newfound status? After all, the US$1 million annual grant gives them the resources to do better work. To sort this out, we can focus only on articles written by the awardees *before* they received the award. Therefore, any citation differences between the two groups couldn't be simply the result of the superior resources offered to awardees. Sure enough, the analysis uncovered a post-appointment citation boost to *earlier* works, offering evidence that in science, the haves are indeed more likely to have more than the have-nots.

This success-breeds-success effect is not limited to HHMI investigators. When a scientist moves from a laureate-to-be to a Nobel laureate, her previously published work – whether of Nobel prize-winning caliber or not – gathers far more attention [78]. Once again, like the case of John Fenn discussed above, a person's previous work doesn't change when she becomes an HHMI investigator or a Nobel laureate. But with new accolades casting warm light on her contribution, attention to her work increases.

Interestingly, though, strictly controlled tests suggest that status has only a modest role on impact, and that role is limited to a short window of time. Consistent with theories of the Matthew effect, a prize has a significantly larger effect when there is uncertainty about article quality, and when prizewinners are of (relatively) low status at the time of the award. Together, these results suggest that while the standard approach to estimating the effect of status on performance is likely to overstate its true influence, prestigious scientists do garner greater recognition for their outputs, offering further support for the Matthew effect.

Box 3.3 Causal evidence for the Matthew effect: Field experiments

Randomized experiments offer the best way to untangle the role of status from individual differences such as talent. We can select two groups – a control and a treatment group – and randomly assign an advantage to some while denying it to others. If success is allocated independent of prior success or status, any discrepancy in the subsequent advantage of recipients over non-recipients can only be attributed to the exogenously allocated early success.

While we can't assign life-altering awards or grants to randomly chosen scientists [79], we *can* explore the phenomenon using experiments carried out in real-world settings where the intervention causes minimal harm. This is what Arnout van de Rijt and his collaborators did in a series of experiments [80, 81]. They randomly selected the most productive Wikipedia contributors within a subset of the top 1 percent of editors and randomly assigned them to one of two groups. Then they gave out "barnstars" to the experimental group – an award used within the community to recognize outstanding editors, while leaving the control group unrecognized. As shown in Fig. 3.2, prior to intervention, the activities of the two groups are indistinguishable, as they were drawn randomly from the same sample of productive editors. Yet once the fake barnstars were bestowed on the experimental group, the awardees exhibited more engagement than their peers in the control group, demonstrating greater sustained productivity and less likelihood of discontinuing their editorial

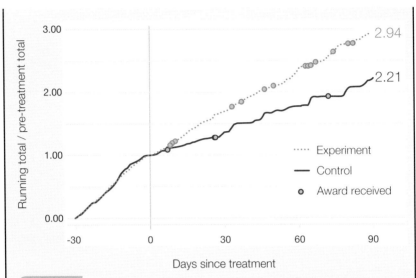

Figure 3.2 **The Matthew effect: Evidence from a field experiment.** Researchers randomly assigned Wikipedia editors into two groups, then awarded barnstars to the experimental group, and did nothing for the control group. Circles show when the editors received additional real awards after the treatment. Twelve subjects in the experimental group received a total of 14 awards, whereas only two subjects in the control condition received a total of three awards. After Restivo and van de Rijt [80].

activity. Indeed, receiving a barnstar increased median productivity by 60 percent compared to the control group. Most importantly, they also went on to win many more *real* barnstars from other editors. These additional awards were not just due to their increased productivity, since within the experimental group, the awarded individuals were not more active than those who received no additional barnstars. The observed success-breeds-success phenomenon was shown to persist across domains of crowdfunding, status, endorsement, and reputation [81], documenting that initial endowments, even if they are arbitrary, can create lasting disparities in individual success, and offering causal evidence for the Matthew effect.

4 AGE AND SCIENTIFIC ACHIEVEMENT

When Elias Zerhouni was named director of the National Institutes of Health (NIH) in 2002, he faced a seemingly intractable crisis: The investigators that his organization funds were aging at an alarming rate. Back in 1980, for example, about 18 percent of all recipients of R01s – NIH's most common research grants for individual investigators – were junior scientists aged 36 or younger. Senior investigators, those 66 years old or above, accounted for less than 1 percent of the grantees. In the 30 years since, however, a remarkable reversal has taken place (Fig. 4.1) [82]: The number of senior investigators increased *ten-fold* by 2010, whereas the percentage of young investigators plummeted dramatically, from 18 percent to 7 percent. In other words, back in 1980, for every senior investigator, the NIH funded 18 junior faculty. By 2010, there were twice as many senior investigators as those just starting out. Zerhouni declared this trend to be "the number one issue in American science" [83].

Why was the director of NIH so worried about this trend? After all, senior researchers have a proven track record, know how to manage projects, understand risks, and can serve as experienced mentors for the next generation of scientists. So, when after a rigorous peer review process, senior principal investigators (PIs) win out over the fresh-faced youngsters, shouldn't we feel reassured that our tax dollars are in safe hands?

To comprehend the threat this demographic shift poses to the productivity and pre-eminence of American science, we need to explore a question that has fascinated researchers for centuries: At what age does a person make their most significant contribution to science?

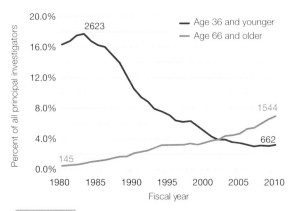

Figure 4.1 **The graying of science.** Changes in the percentage of NIH R01 grant recipients, aged 36 and younger and aged 66 and older, 1980–2010. After Alberts et al. [82].

4.1 When Do Scientists Do their Greatest Work?

The earliest investigation into the link between a person's age and exceptional accomplishment dates back to 1874, when George M. Beard estimated that peak performance in science and the creative arts occurred between the ages of 35 and 40 [84]. Subsequently, Harvey C. Lehman devoted around three decades to the subject, summarizing his findings in *Age and Achievement*, a book published in 1953 [85]. Since then, dozens of studies have explored the role of age in a wide range of creative domains, revealing a remarkably robust pattern: No matter what creative domain we look at or how we define achievement, one's best work tends to occur around mid-career, or between 30 to 40 years of age [2, 66, 85–87].

Figure 4.2 shows the age distribution of signature achievements, capturing Nobel prizewinners and great technological innovators of the twentieth century [88]. The figure conveys three key messages:

(1) There is a large variance when it comes to age. While there are many great innovations by individuals in their 30s (42%), a high fraction contributed in their 40s (30%), and some 14 percent had their breakthrough beyond the age of 50.

(2) There are no great achievers younger than 19. While Einstein had his *annus mirabilis* at the tender age of 26, and Newton's *annus mirabilis* came even earlier, at the age of 23, the Einsteins and Newtons of the world are actually rare, because only 7 percent of

41 / Age and Scientific Achievement

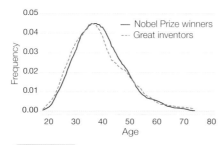

Figure 4.2 **Age distribution of great innovation.** The plot shows the age of innovators at the moment of their great innovation, combining all twentieth-century observations. After Jones [88].

the sample have accomplished their great achievement at or before the age of 26.

(3) The Nobel laureates and the inventors come from two independent data sources, with only 7 percent overlap between the two lists. Yet, the age distributions of these two samples are remarkably similar.

Thus, Fig. 4.2 demonstrates that scientific performance peaks in middle age [2, 66, 85–87]. The life cycle of a scientist often begins with a learning period, absent of major creative outputs. This is followed by a rapid rise in creative output that peaks in the late 30s or 40s and ends with a slow decline as he advances through his later years. These patterns are remarkably universal. Researchers have explored them in a variety of ways, identifying important scientists by their Nobel Prizes, by their listings in encyclopedias, and by their membership in elite groups like the Royal Society or the National Academies. No matter how you slice the data, the patterns observed in Fig. 4.2 remain essentially the same, raising two questions: Why does creativity take off during our 20s and early 30s? And why does it decline in later life?

4.2 The Life Cycle of a Scientist

4.2.1 The Early Life Cycle

A remarkable feature of a scientific career is the lack of contributions in the beginning of life [89]. Simply put, no 18 year old has managed to produce anything worthy of a Nobel. The early life cycle coincides with schooling, suggesting that the need for education may be responsible for the absence of exceptional output in early age. This is also consistent with theories of creativity, where "creativity" is often defined as an ability to identify interesting combinations of existing knowledge [90–92]. Indeed, if we see knowledge as Lego pieces, and new inventions and ideas hinge on figuring out novel ways to combine these pieces, then we first need to accumulate enough pieces before we can start building anything meaningful.

Empirical evidence using World Wars I and II as natural experiments [88], supports the idea that training is responsible for the lack of discovery and invention in the early life cycle. Indeed, war created interruptions that affected the training phase of careers. When Nobel laureates experienced either World War I or II between the ages of 20 and 25, their probability of creating innovative work between the ages of 25 and 30 was substantially reduced, even though the wars were over. This means, scientists do not magically start innovating at a certain age; rather, interruptions during their training phase must be made up. This finding is consistent with the much-discussed "10,000-hours hypothesis" in psychology, which estimates that roughly 10 years of deliberate practice are needed to achieve excellence across broad domains of human performance [93–95].

4.2.2 The Middle and Late Life Cycle

Once a scientist completes the training phase, a productive career may blossom, as evidenced by the prominent peak in Fig. 4.2. Yet, Fig. 4.2 also shows that the likelihood of a scientific breakthrough decreases following this peak around middle age. Compared to the rapid rise in the early life cycle, this decline is relatively slower.

There are many potential explanations for this decline, from the obsolescence of skills to degradation in health. But none of them offer a satisfactory explanation for why the decline occurs so early, usually before a scientist reaches 50 – long before he is considered old, biologically. For this reason, alternative explanations that consider the realities of the research process stand out as more plausible, including family responsibilities and increasing administrative duties. In other words, older scientists may well have the capacity to produce great science but have limited time to do so. Instead, they're running labs, applying for grants, reviewing papers and deciding tenure cases. Interestingly, scientific careers are sometimes characterized by a "second peak" prior to retirement, which can be interpreted as a rush to get remaining ideas and unpublished research in press [89, 96–98].

These findings help us return to the challenge facing American science (Fig. 4.1): While senior researchers have more experience, scientists tend to produce their best work at a younger age, suggesting that a preference for funding only older researchers could stifle innovation. For example, from 1980 to 2010, 96 scientists won a Nobel Prize in medicine or chemistry for work supported by NIH. Yet their prizewinning research was conducted at an average age of 41 [99] – a full year younger than the average age of a NIH investigator starting out today.

4.3 What Determines the Timing of Our Peak?

What determines when our creativity peaks? Many believe that a scientist's field is a key factor, as some fields are "young" and others are "old." Indeed, the prevailing hypothesis is that individuals working in disciplines that focus on deduction and intuition tend to have their great achievements at younger ages, explaining why peak performance comes earlier in disciplines like math or physics than in medicine [85, 100–102].

But this stereotype – of precocious physicists and wizened medical doctors – is increasingly contested. For example, researchers tabulated peak age in different fields by surveying the literature. Upon putting all these numbers together, they found that the peak age was all over the map [89], lacking obvious so-called young or old fields. Even the canonical example of physics vs. medicine doesn't seem to hold. For example, while it's true that physics Nobel laureates during the 1920s and 1930s were younger than scientists in all other fields, since 1985, physics laureates have actually been *older* when they made their prizewinning contribution compared to other fields [103].

If field is not a determining factor, then what is? As we discuss next, two convincing theories have emerged: the burden of knowledge and the nature of work.

4.3.1 Burden of Knowledge

A simple re-examination of Fig. 4.2 reveals an interesting pattern [88]. If we take the datasets of Nobel prizewinners and great

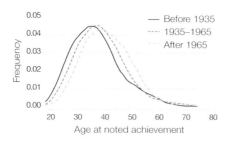

Figure 4.3 Shifts in the age distribution of great innovation. Data includes Nobel prizewinners and great inventors. The twentieth century is divided into three chronological periods: 1900–1935, 1935–1965, and 1965 to present. After Jones [88].

inventors, and divide the twentieth century into three consecutive chronological periods, we observe a systematic shift (Fig. 4.3): The peak age of great minds has increased over time. Indeed, in the early days of the Nobel prize, two thirds of the recipients did their prize-winning work by age 40, and 20 percent earned the distinction before they turned 30. Lawrence Bragg, the youngest Nobel laureate in physics, received his award at the astonishing age of 25. Today, a 25-year-old physicist has probably only recently decided on which PhD program to enroll at to begin her training. And since 1980, the mean age for Nobel winning achievement in physics has shifted to 48. Overall, as the twentieth century progressed, the great achievements by both Nobel laureates and inventors have occurred at later and later ages, with the mean peak age rising by about six years in total.

There are two plausible explanations for this shift. The first hypothesis is that the life cycle of innovation has changed, so that great minds now innovate at a later stage of their career. This could be due to the extended length of time required for education, which delays the onset of active innovative careers. The second hypothesis reasons that the upward age trend simply reflects general demographic shifts. In other words, if everyone in the world is getting older on average, science as a profession should be no exception.

Yet, even after controlling for demographic effects, substantial shifts in peak age remain unexplained, especially the delays in the early life cycle, when innovators began their active careers [88]. For example, while at the start of the twentieth century a scientist became "research active" at age 23, by the end of the century this moment had shifted to age 31.

To understand the origin of the observed shift, Jones proposed the "burden of knowledge" theory [88, 104, 105]. First, innovators must undertake sufficient education to reach the frontier of knowledge. Second, because science has been growing exponentially, the amount of knowledge one has to master to reach that frontier increases with time. This theory offers a fresh perspective of Newton's famous remark about "standing on the shoulders of giants": In order to stand on a giant's shoulders, one must first climb up his back. But the greater the body of knowledge, the longer the climb.

Clearly, education is an important prerequisite for innovation [88]. Yet if scientists tend to innovate when they are young, then each additional minute they spend in training is one less minute they can spend pushing science forward, potentially reducing the total career output of individual scientists. And if individuals have less time to innovate, society as a whole innovates less [89]. A back-of-the-envelope calculation suggests that a typical R&D worker contributes approximately only 30 percent of his time to aggregate productivity gains as he did at the beginning of the twentieth century – a shift that can be attributed to the burden of knowledge. In other words, an increase in the age of peak performance potentially contributes to the well-documented decline in per capita output of R&D workers in terms of both patent counts and productivity growth [106].

Box 4.1 Nobel Prize threatened

It's not only discovery that's delayed these days – so is recognition [107]. Indeed, there is a steadily increasing gap in time between when a scientist makes a discovery worthy of a Nobel and when she is awarded for it. Before 1940, only 11 percent of physics, 15 percent of chemistry, and 24 percent of physiology or medicine Nobel Prize recipients had to wait more than 20 years for their recognition after the original discovery. Since 1985, a two-decade delay has affected 60 percent, 52 percent, and 45 percent of the awards, respectively. The increasing interval between discovery and its formal recognition is approximated by an exponential curve, projecting that by the end of this century, the prizewinners' average age for receiving the award is likely to exceed their projected life expectancy. In other words, most candidates will not live long enough to attend their Nobel ceremonies. Given that the Nobel Prize cannot be awarded posthumously, this lag threatens to undermine science's most venerable institution.

4.3.2 Experimental vs. Conceptual Innovators

By the time Heisenberg entered the University of Munich in 1920, it was clear that the leading theory of the atom developed by Bohr, Sommerfeld, and others, while successful in certain domains, had encountered fundamental challenges. Heisenberg earned his PhD in Munich with Arnord Sommerfeld in 1923 and moved on to work with Max Born in Göttingen, where he developed matrix mechanics in 1925. The following year he became Niels Bohr's assistant in Copenhagen, where, at the age of 25, he proposed the uncertainty principle. In the past century, historians have repeatedly documented his "extraordinary abilities: complete command of the mathematical apparatus and daring physical insight" – to borrow from Sommerfeld, his PhD advisor [108].

What is less known, however, is that he nearly failed his PhD defense. On July 23, 1923, the 21-year-old Heisenberg appeared before four professors in the University of Munich. He easily handled Sommerfeld's questions and those relating to mathematics, but he stumbled with astronomy, and flunked badly in experimental physics [108]. When all was said and done, he passed with a C-equivalent grade. When he showed up the next morning at Max Born's office – who had already hired him as his teaching assistant for the coming year – Heisenberg said sheepishly, "I wonder if you still want to have me."

Why, you might ask, did an astonishingly brilliant physicist fail an exam covering the basics of his field? Perhaps because of the kind of innovator he was. There are, broadly speaking, two polar extremes in scholarly creativity: conceptual and experimental [109]. *Experimental innovators* work inductively, accumulating knowledge from experience. This type of work relies heavily on the works of others and tends to be more empirical. Heisenberg was, on the other hand, a *conceptual innovator*, who worked deductively, applying abstract principles. Consequently, his work tended to be more theoretical and derived from a priori logic. This distinction between experimental and conceptual creativity suggests that the nature of the work a scientist does affects when she peaks: Conceptual innovators tend to do their most important work earlier in their careers than their experimental colleagues.

To explore the distinct life cycles of the two types of scientists, researchers rated Nobel-winning economists as either conceptual or experimental based on the nature of their discovery, and examined the age when each group published their best work [109]. The difference

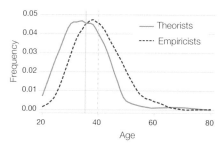

Figure 4.4 Theorists vs. empiricists among Nobel laureates. This figure separates profiles for Nobel laureates, whose prizewinning work was theoretical compared to those whose prizewinning work was empirical, showing clear differences in life cycle of creativity between the two. After Jones [89].

was stark: Conceptual laureates made their most important contributions to science at the average age of 35.8. Compare that to the average age of 56 for experimental laureates – a staggering difference of 20.2 years. Indeed, 75 percent of conceptual Nobel laureates in economics published their best work within the first 10 years of their career, while *none* of the experimental laureates managed to do so.

Another way of looking at the nature of work is to separate it into the categories of empirical or theoretical. Note that this distinction is not equivalent with conceptual vs. experimental. Indeed, while conceptual innovations tend to be more theoretical, a conceptual innovator can certainly be rooted in empirical work as well. Likewise, experimental innovators can also make theoretical contributions. Nevertheless, when Nobel laureates were separated into either empirical or theoretical categories, a similar pattern emerged (Fig. 4.4) [89]: the empirical researchers did their prizewinning work 4.6 years later (at 39.9 years of age compared to 35.3 years) than those doing theoretical work.

There are many reasons why conceptual work tends to happen early and experimental work tends to happen later in a career [89]. First, conceptual innovators like Heisenberg do not need to accumulate large amounts of experience in order to make their contributions. By contrast, Charles Darwin, who did experimental work, needed time to accumulate the evidence necessary to develop his theories. Sure, he was a brilliant naturalist, but his work would have gone nowhere if he hadn't boarded the HMS *Beagle* to voyage around the world. Collecting all the pieces of evidence and drawing connections among them took time. Indeed, the trip itself took five years. That means it took Darwin a half-decade of travel to even have the raw materials for writing *The Voyage of the Beagle*.

Second, some of the most important conceptual work involves radical departures from existing paradigms, and it may be easier to disrupt a paradigm shortly after initial exposure to it, before an individual has produced a large body of work that rests upon the prevailing wisdom. Thus, while experience benefits experimental innovators, newness to a field benefits conceptual innovators. Indeed, nearly failing his PhD exam did not prevent Heisenberg from doing groundbreaking work. To a certain degree, it might have even helped. In other words, knowing too much might even "kill" your creativity [110, 111].

In some sense, conceptual and experimental innovators are like a Kölsch beer and a vintage port. The former is best served fresh, whereas the latter ages with more pleasing mouthfeel and softening of tannins. This contrasting approach to the creative process translates to fields beyond science. Among important conceptual innovators in the modern era, Albert Einstein, Pablo Picasso, Andy Warhol, and Bob Dylan all made their landmark contributions between the ages of 24 and 36, whereas the great experimental innovators, such as Charles Darwin, Mark Twain, Paul Cézanne, Frank Lloyd Wright, and Robert Frost, made their greatest contributions between the ages of 48 and 76 [112, 113]. The nature of work is such a strong predictor of when an innovator peaks, that sometimes you don't even need data to see its effect. Take for example art. In Fig. 4.5, we put side-by-side works by

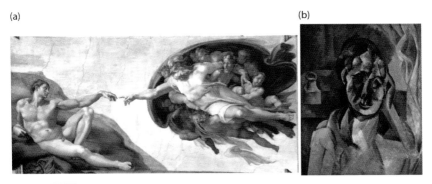

Figure 4.5   **Michelangelo vs. Picasso.** Two famous paintings side by side. (a) *The Creation of Adam*, by Michelangelo (1511–1512). (b) *Woman with Mustard Pot*, by Picasso (1910). Can you guess which of these two artists' career peaked early and which is the one that required longer time to mature?

Michelangelo and by Picasso. Can you guess who is the Kölsch beer and who is the vintage port?

> **Box 4.2 Age and adoption of new ideas: Planck's principle**
>
> Widely held is the idea that younger researchers are more likely to make radical departures from convention. Max Planck vividly put it this way: "A new scientific truth does not triumph by convincing its opponents and making them see the light, but rather because its opponents eventually die, and a new generation grows up that is familiar with it." This school of thought argues age as an important factor driving scientific progress. In other words, if young and old scientists have different affinities for accepting new ideas, then scientific progress will have to wait for older scientists to fade in relevance, before the discipline can move forward. Yet by contrast, another school of thought argues that new ideas triumph when they are supported by more empirical evidence than the competing hypotheses. Therefore in science, reason, argument, and evidence are all that matter, suggesting that the age factors may not be as important as we thought. So is what Planck said true for science?
>
> Although people have long suspected that younger researchers are more willing to make radical departures from convention, there has traditionally been limited evidence supporting Planck's principle. For example, Hull and colleagues found that older scientists were just as quick to accept Darwin's theory of evolution as younger scientists [114]. This finding supports the idea that scientific progress is self-correcting, guided only by truth. Yet a study focusing on 452 prominent life scientists who passed away suddenly while at their peak of their career, offers empirical support for Planck's principle [115]. After the unexpected death of a prominent scientist, her frequent collaborators – the junior researchers who coauthored papers with her – experience a sudden drop in productivity. At the same time, there is a marked increase in published work by newcomers to the field, and these contributions are disproportionately likely to be highly cited. They are also more likely to be authored by young outsiders to the field.
>
> These results are also consistent with a study that investigates the link between age and the adoption of new ideas. By measuring new words used in a paper as a proxy for new ideas, researchers found that young researchers are much more likely than older scientists to tackle exciting, innovative topics – and when they do, the resulting work has a higher impact [111]. These results document a "Goliath's shadow" effect in science: Outsiders are reluctant to challenge the leading thinker within a field while he is still alive, but stream in and take over leadership once he has passed away.

Regardless of whether the scientists applying for NIH funding were conceptual or experimental innovators, Elias Zerhouni was right to be concerned: One needs the funds to innovate immediately, the other to collect the evidence that would facilitate her breakthrough decades later. Overall, this chapter unveils an intimate link between age and scientific achievement, showing that a scientist's performance peaks relatively early, followed by a rapid, if brutal decline. Once we pass the peak, hope for a breakthrough starts to dim.

Or so it seemed anyway. While the research we shared here is solid, older scientists can maintain their hope. Because, as we show in the next chapter, there's more to the story.

5 RANDOM IMPACT RULE

Much of the research on age and achievement discussed so far has one thing in common: The focus is on famous scientists, those that we laud as geniuses. So, are their conclusions relevant to us mere mortals?

This long-standing focus on prominent scientists makes sense methodologically: most of the existing knowledge in this space was tabulated by hand, scrawling down the dates of major works and estimating the scientist's age when he completed them, and sometimes locating the evidence deep in the library stacks. Equally important, information on prominent scientists is easier to come by, as they were collected in biographies and laudations.

Even today, as computers have significantly eased our ability to collect and organize data, it remains a difficult task to study individual careers, given the name disambiguation challenges discussed in Chapter 1 (Box 1.1). Yet, thanks to advances in data mining and machine learning, which uses information from research topic to the author's institution to citation patterns, the accuracy of name disambiguation has improved considerably in the past decade. As a result, career histories of individual scientists – not just geniuses, but also everyday Joes and Janes slogging it out in the field – are becoming available to researchers on a much larger scale. These advances offer new opportunities. In fact, as we will see in this chapter, the data did not just test and validate existing theories and frameworks. It upended the way we think about individual careers altogether.

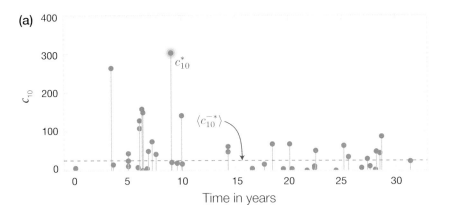

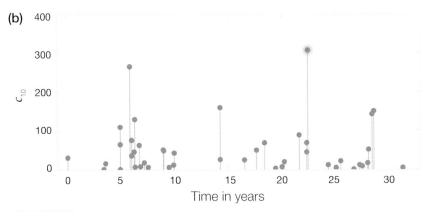

Figure 5.1 Publication history of Kenneth G. Wilson. (a) The horizontal axis indicates the number of years after Wilson's first publication and each vertical line corresponds to a publication. The height of each line corresponds to c_{10}, i.e., the number of citations the paper received after 10 years. Wilson's highest impact paper was published in 1974, 9 years after his first publication; it is the 17th of his 48 papers, hence $t^* = 9$, $N^* = 17$, $N = 48$. (b) Shuffled career of Wilson, where we keep the locations of the pins, but swap the impact of each paper with another one, thereby breaking the temporal ordering of when best work occurs within a career. After Sinatra et al. [116].

5.1 Life Shuffled

Figure 5.1a shows the scientific career of Kenneth G. Wilson, the 1982 Nobel laureate in physics. We treat his first publication as the start of his academic timeline, and every time he publishes another paper, we add one more pin at the corresponding time point (academic age) in his career. The heights of each pin shows the paper's impact, approximated by the number of citations the paper received 10 years after its publication.

This "drop pin" view allows us to represent the career of every scientist in the same way. When repeated for tens of thousands of scientists from all kinds of disciplines, it can help us answer a simple question that has long been elusive, despite the extensive literature on geniuses: When does an *ordinary* scientist publish her best work?

Initial attempts to answer the question focused on the career histories of 2,856 physicists whose publication records span at least 20 years [116], extracted from the publication records of 236,884 physicists publishing in *Physical Review*. The dataset allowed us to identify each physicist's personal hit, i.e., the paper that collected more citations than any other paper he published. To understand when a scientist publishes his highest impact work, we measure t^*, the academic age of the scientist when the hit paper was published. Hence t^* marks, for example, the *Penicillium chrysogenum* paper for Alexander Fleming, and the radioactivity paper for Marie Curie. But it could also denote the far less cited, yet nonetheless personal best paper by your colleague in the office next door.

Figure 5.2 plots $P(t^*)$, the probability that the highest impact paper is published t^* years after a scientist's first publication. The high

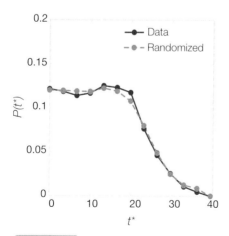

Figure 5.2 The random impact rule. Distribution of the publication time t^* of the highest impact paper in scientists' career (black circles) and for randomized impact careers (black circles). The lack of differences between the two curves indicates that impact is random within a scientist's sequence of publication. After Sinatra et al. [116].

$P(t^*)$ between 0 and 20 years indicates that most physicists publish their highest impact paper in early or mid-career, followed by a significant drop in $P(t^*)$ beyond this period. It shows that once scientists pass their mid-career, the possibility of creating breakthrough work becomes depressingly unlikely.

Yet, upon a closer examination, it turns out that the interpretation of this curve is not as straightforward as it seems at first. To see this, we next ask how would the same plot look, if the timing of the highest impact work is driven entirely by chance?

Imagine for a second that creativity in a career is purely random. To find out how such a random career would look, we take two pins at random and swap them, and we do this over and over, thousands of times. In doing so, we arrive at a shuffled version of each scientist's career (Fig. 5.1b). Where does the shuffled career differ from the real one? The net productivity of the individual has not changed. The total impact of these papers didn't change either, since we did not alter the size of the pins. Nor did we change the timing of the publications. The only thing that *did* change is the order in which these papers were published. Imagine a lifetime of papers as the deck of cards you're dealt, and your highest impact paper is your ace of diamonds, then we just shuffled the order of those cards, including the ace. Your ace of diamonds now can appear anywhere – top, middle, bottom.

Next, we measured $P(t^*)$ for the shuffled careers and plotted the randomized version of $P(t^*)$ together with the real one in the same figure. To our surprise, the two curves in Fig. 5.2 are right on top of each other. In other words, the timing of the best works in the randomly shuffled careers is indistinguishable from the original data. What does this mean?

5.2 The Random Impact Rule

The fact that the two distributions in Fig. 5.2 are the same indicates that variations in $P(t^*)$ are *fully explained by changes in productivity* throughout a career. Indeed, the randomized curve measures the variations in productivity during a scientist's career. It shows that in this sample of scientists, productivity has a peak at year 15 of a career, and it drops rapidly after year 20. This means that young scientists have a disproportionate number of breakthroughs early in their career not because youth and creativity are intertwined, but simply because they're in their most productive period. In other words, when we adjust for productivity, high impact work will occur randomly over the course of a career. We call this the *random impact rule* [116].

Staying with the card analogy, imagine you draw one card at a time from your deck, but with varying frequency. You start furiously in the beginning of your career, drawing one card after another, rapid-fire, excited by your work and hoping to find that ace. This rapid-fire period is then followed by a gradual decline, where you slow down how frequently you reach out to the deck. Now, if the deck is well shuffled beforehand, and you draw a lot more cards during the first 20 years than during your later period, when will you most likely encounter the ace? During the first 20 years, of course. In other words, the first two decades of your career are not more creative than the later 20 years. You draw an ace early in your career simply because you try harder.

To more directly test the random impact rule, we can look at where the ace appears within the deck. For that we calculate the position of the highest impact paper N^* in the sequence of N publications of a scientist. Then, we measure $P(N^*/N)$, i.e. the probability that the most cited work is published early (small N^*/N) or late ($N^*/N \approx 1$) in the sequence. If the random impact rule holds true, $P(N^*/N)$ should follow a uniform distribution: That is, with equal probability we should find the hit paper at any positions of N^*/N. In technical terms, this means that the cumulative distribution $P^>(N^*/N)$ must decrease as a straight line, following $(N^*/N)^{-1}$. As Fig. 5.3a shows, the data follows exactly what the random impact rule predicts.

But why stop at the highest impact paper? What about the second highest? How about the third? As you may have guessed, the same pattern emerges (Fig. 5.3a). The cumulative distribution follows a clear straight line. That is, the big break of your career can come at any time, and this pattern is not limited to your highest impact work – your other important works are equally random [117]. And this random impact rule holds not only in scientific careers, but also in careers across different creative domains, like artists and film directors [117] (Fig. 5.3).

The random impact rule has deep roots in the literature, dating back to the 1970s work by Simonton, who proposed the "constant-probability-of-success" model [2, 118–121]. Researchers have long suspected that the same rule holds for the arts, such as literary and musical creativity [118]. But it took more than 40 years to collect the appropriate datasets to have it formally tested.

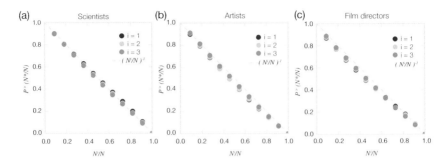

Figure 5.3 Random impact rules across creative domains. Cumulative distribution $P^>(N^*/N)$, where N^*/N denotes the order N^* of the highest impact paper in a career, varying between $1/N$ and 1. The cumulative distribution of N^*/N is a straight line with slope -1, indicating that N^* has the same probability to occur anywhere in the sequence of works one produces. Figure shows $P^>(N^*/N)$ for careers of (a) 20,040 scientists, (b) 3,480 artists, and (c) 6,233 film directors [112]. For each creative individual, we identified their top three highest impact works (publications, artworks, movies) and measured their relative position within their career (citations, auction prices, ratings recorded in the Internet Movie Database (IMDb). The panels demonstrate that, across all three careers, the timing of each of the three highest impact works is random within the sequence of works one produces. After Liu et al. [117].

The random impact rule changes our understanding of when breakthroughs happen in an individual's career. Indeed, decades of research have documented that major discoveries often come early in a scientist's career. This has led to the view that creativity equals youth, a myth deeply ingrained in the popular culture. The random impact rule helps us decouple age and creativity. It tells us that the chance of a breakthrough is completely random within the sequence of works produced in a career. To be precise, every project we do has the same chance of becoming our personal best. What is not random, however, is productivity: younger researchers are more eager to try over and over, putting papers out one after another. If the impact is random within an individual's range of projects, then it is inevitable that, statistically speaking, it will happen early on in an individual's career, when productivity is high.

The random impact rule thus offers a new perspective on the role of productivity: it tells us that trying over and over is important for making that long-awaited breakthrough. Indeed, for those that keep trying, the breakthrough may not be as elusive. A wonderful example is offered by John Fenn (Fig. 5.5). He identified a novel electrospray ion

Box 5.1 Lost winners

The results of Chapter 4 showed that the prizewinning works by Nobel laureates tend to occur early within their career. By contrast, this chapter shows that ordinary scientific careers are governed by the random impact rule. Does the random impact rule apply to Nobel careers as well [122]? To find out we measured the positions of the prizewinning work and most-cited work within the sequence of papers published before being awarded the Nobel Prize (51.74 percent of the most-cited papers were also the prizewinning papers), finding that both tend to occur early within the sequence of papers (Fig. 5.4). This suggests that compared with ordinary scientists, Nobel laureates tend to do their best work disproportionately early in their careers.

Yet, there is a selection effect we must confront – since the Nobel Prize in science has never been awarded posthumously, those who produced groundbreaking works early were more likely to be recognized. To test this hypothesis, we removed prizewinning papers, which are subject to this selection bias, and measured the timing of each of the three remaining highest impact works for Nobel laureates, finding that they are all distributed randomly within their career (Fig. 5.4). This means, apart from the prizewinning work, all other important works in a Nobel-winner's career follow the random impact rule. One implication of this selection bias is the existence of "lost winners," scientists whose deserving works were not recognized by a Nobel simply because their work came late in their career.

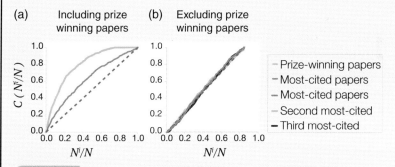

Figure 5.4 Career patterns of Nobel laureates. (a) The cumulative distributions of relative positions (N^i/N) of the prizewinning papers and the most-cited papers within the sequence of all papers before being awarded the Nobel prize. The dashed line indicates the predictions of the random impact rule. (b) To eliminate potential selection bias in the timing of the prizewinning work, we removed prizewinning papers and calculated the relative position of the top three most-cited papers among all the papers published before the award, finding that these papers follow the random impact rule. After Li et al. [122].

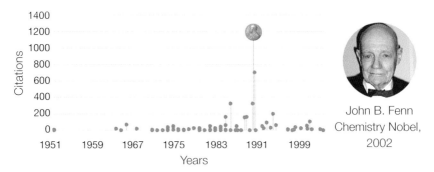

Figure 5.5 The academic career of John B. Fenn, the 2002 Nobel laureate in chemistry.

source at the very end of his official academic career, just as he was forcefully retired by Yale University. Undeterred, he left Yale, took a new faculty position at Virginia Commonwealth University, and continued his investigation, which eventually led to the discovery of electrospray ionization, for which, 15 years later, he received his Nobel Prize. Overall, his example, together with the random impact rule, tells us that for those who do not let their productivity drop in the later part of their career, impact may not wane.

While the random impact rule deepens our understanding of patterns underlying scientific careers, it also raises a new puzzle: If the timing of the biggest hit is random, what is not random in a career?

Box 5.2 The old myth of the young entrepreneur

The youth=creative dogma is not limited to science – it is deeply ingrained in the world of entrepreneurship as well. Indeed, the average age of the winner of the TechCrunch awards in Silicon Valley is 31, and those named "Top Entrepreneurs" by *Inc.* and *Entrepreneur* magazines have an average age of 29. Similarly, the average age of the founders backed by one of the best-known VC firms, Sequoia, is 33, and those supported by Matrix Ventures were on average 36. But does youth imply success in Silicon Valley?

By combining tax-filing data, US Census information, and other federal datasets, researchers compiled a list of 2.7 million company founders [123]. Their analysis revealed that, contrary to popular thinking, the best entrepreneurs tend to be middle-aged. Among the fastest-growing new tech companies, the average founder was 45 at the time of founding.

Furthermore, a 50-year-old entrepreneur is nearly twice as likely to have a runaway success as a 30 year old.

These results show that entrepreneurial performance rises sharply with age. Indeed, if you were faced with two entrepreneurs and knew nothing about them besides their age, contrary to the prevailing wisdom, you would do better, on average, by betting on the older one (Fig. 5.6).

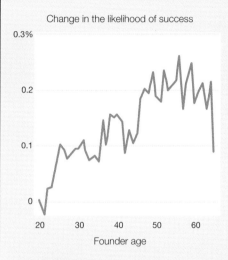

Figure 5.6 **Older entrepreneurs are more likely to succeed.** The probability of startup success rises with age, at least until the late 50s. The y-axis represents the ordinary least squares regression coefficient for age variables, measuring change in the likelihood of extreme success relative to a 20-year-old founder. Here "extreme startup success" is defined as the top 0.1 percent of startups in employment growth over five years [123]. After Azoulay et al. [124].

6 THE *Q*-FACTOR

What distinguishes successful scientists from their peers? Our ability to identify the roots of a successful career is limited by the fact that productivity, impact, and luck are intimately intertwined. Indeed, if breakthroughs in a career occur at random, what is the relevance of luck, talent, or hard work? Can we at all separate the innate talent and ability of a scientist from his luck? To understand these questions, let's start with a thought experiment: How likely is it that Einstein's exceptional output may emerge purely by chance?

6.1 Pure Coincidence

Given an infinite length of time, a chimpanzee punching at random on a typewriter will surely type out a Shakespeare play. So, with a sufficiently large number of scientists in the world, shouldn't we expect that chance alone will inevitably lead to someone as impactful as Einstein?

The random impact rule discussed in Chapter 5 allows us to build a "null" model of scientific careers to answer this question. Let us assume for the moment that for a scientist a publication is the equivalent of drawing a lottery ticket. In other words, we assume that talent plays no role, allowing us to ask, what does a career driven by chance alone look like?

In a random career the impact of each paper is determined solely by luck. This means that we simply pick a random number from a given impact distribution and assign it to the paper published by our scientist. This procedure allows us to generate a collection of "synthetic" careers which are purely driven by chance. For convenience, we will call this procedure the random model, or the *R*-model.

In some ways, these random careers may appear quite similar to the real ones. For example, they will show individual differences in career impact, as some scientists are inevitably luckier than others when they select their random numbers. And each career will obey the random impact rule as well: Since the impact of every paper is chosen randomly, the highest impact work will be random in the sequence of papers published by each scientist. But will these random scientists differ from the real ones?

If each paper's impact is drawn randomly from the same impact distribution, a more productive scientist will draw more tickets, hence will more likely stumble upon a high impact publication. In other words, the *R*-model predicts that more productive scientists are more likely to make breakthroughs. To test this effect, we measured how the impact of a scientist's most-cited paper, $\langle c_{10}^* \rangle$, depends on her productivity, N. We find that indeed, the more papers a scientist publishes, the higher is the impact of her highest impact paper. Yet, not high enough: The measurements indicate that in real careers the impact of the highest impact work increases faster with N than the *R*-model predicts (Fig. 6.1). In other words, as scientists become more productive, their home-run papers are much more impactful than what we would expect if impact is like a lottery ticket, randomly assigned. This indicates that something is missing from our null model. It's not hard to guess what: scientists inherently differ from each other, in either talent or ability or other characteristics relevant to producing high-impact works. This suggests that highly productive scientists don't stand out in productivity alone – they possess something that the low-productivity scientists do not, which helps them publish higher-impact work. Next, we adjust the model to account for the fact that not all scientists are alike.

6.2   The *Q*-Model

Each scientific project starts with an idea. A light bulb goes off, prompting the scientist to think, "I'm curious if that idea might work." But it is hard to evaluate the inherent importance or novelty of the idea in advance. Not knowing what its true value is, let's assume that the idea has some random value r. Some ideas are of incremental

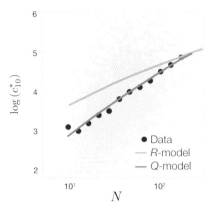

Figure 6.1 **Scientific careers are not random.** Scatter plot shows the citation of the highest impact paper, c^*_{10} vs. the number of publications N during a scientist's career. Each gray dot corresponds to a scientist. The circles are the logarithmic binning of the scattered data. The cyan curve represents the prediction of the R-model, which shows systematic deviation from data. The red curve corresponds to the analytical prediction of the Q-model. After Sinatra et al. [116].

importance, only interesting to a limited number of people in our immediate field, which would correspond to a modest r. Occasionally, though, we stumble upon an idea that could be transformative – that is, if it works out. The better the idea – the larger its r value – the more likely it is to have high impact.

But having a good idea isn't enough. The ultimate impact of the project depends on a scientist's ability to turn that idea into a truly impactful product. One may begin with a terrific idea, but lacking the necessary expertise, experience, resources, or thoroughness to bring out its full potential, the end result will suffer. Yet, turning that idea into a discovery requires an ability that varies from person to person. So, let us assign a parameter Q to characterize an individual's ability to turn a random idea r into a discovery of a given impact.

In other words, let us assume that the impact of each paper we publish, c_{10}, is determined by two factors: luck (r) and the Q_i parameter unique to individual i. The combination of the two could invoke some complicated functions. To keep it simple, we assume a simple linear function, writing

$$c_{10} = rQ_i. \tag{6.1}$$

Equation (6.1) has multiple assumptions behind it:

(1) When we start a new project, we pick a random idea r from a pool of possibilities. Scientist may pick their r from the same distribution $P(r)$, as we all have access to the same literature and hence the same pool of knowledge. Or each scientist may pick her r from her own $P(r)$, as some scientists are better at picking good ideas than others.

(2) Scientists differ in their Q parameter, which characterizes the ability of an individual to take the same idea but turn it into works with variable impacts. If a scientist with a low Q-factor comes across an idea with huge r value, despite the idea's inherent potential, the project's impact will be mediocre, since the resulting product rQ is diminished by the scientist's limited Q. On the other hand, a high Q scientist may also put out weak or mediocre works, if the idea he started with was poor (small r). Truly high-impact papers are those perfect-storm instances when a high Q individual stumbles upon a brilliant idea (high r). In other words, the model assumes that the ultimate impact of a paper is the product of two factors: a yet-to-be-realized potential of an idea and a scientist's ability to realize it.

(3) Finally, productivity matters: Scientists with high N, even with the same Q and $P(r)$, have more chances to stumble across a high r project, turning it into a paper with a high impact (c_{10}).

The problem is that none of these factors are expected to be independent of each other: high Q individuals may also have a talent at recognizing high-potential projects, hence their $P(r)$ distribution could be skewed towards higher r values. And those who publish higher-impact papers may also have more resources to publish more papers, hence will have higher productivity. In other words, the outcome of the model (6.1) is determined by the joint probability $P(r, Q, N)$, with unknown correlations between r, Q, and N. To understand what real careers look like, one needs to measure the correlations between the three parameters. This measurement led to the covariance matrix [116]:

$$\Sigma = \begin{pmatrix} \sigma_r^2 & \sigma_{r,Q} & \sigma_{r,N} \\ \sigma_{r,Q} & \sigma_Q^2 & \sigma_{Q,N} \\ \sigma_{r,N} & \sigma_{Q,N} & \sigma_N^2 \end{pmatrix} = \begin{pmatrix} 0.93 & 0.00 & 0.00 \\ 0.00 & 0.21 & 0.09 \\ 0.00 & 0.09 & 0.33 \end{pmatrix}, \quad (6.2)$$

which makes two unexpected predictions about individual careers:

(1) $\sigma_{r,N} = \sigma_{r,Q} \cong 0$ indicates that the value of an initial idea (r) is largely independent of a scientist's productivity N or her Q-factor. Therefore, scientists source their ideas randomly from a $P(r)$ distribution that is the same for all individuals, capturing a universal – that is, scientist-independent – luck component behind impact.
(2) The nonzero $\sigma_{Q,N}$ indicates that the hidden parameter Q and productivity N do correlate with each other, but its small value also shows that high Q is only slightly associated with higher productivity.

The lack of correlations between the idea value r and (Q, N) allows us to analytically calculate how the highest impact paper c_{10}^* is expected to change with productivity N. As shown in Fig. 6.1, the prediction of the Q-model is now in excellent agreement with the data, indicating that the hidden Q-factor and variations in the productivity N can explain the empirically observed impact differences between scientists, correcting the shortcomings of the R-model.

The fact that we need other parameters besides luck to characterize a scientist's career impact makes sense. It's not hard to imagine that there are differences between individual scientists, which need to be accounted for to have an accurate description of real careers. What is surprising, however, is that we appear to need only *one* additional parameter besides luck. Incorporating the Q-factor *alone* seems sufficient to explain the impact differences among scientists.

What exactly does the Q-model offer that the R-model misses? The R-model's failure tells us that a successful career isn't built on chance alone. Indeed, the Q-factor pinpoints a critical characteristic of a career: Great scientists are *consistently* great across their projects. True, each scientist probably has one important paper they are known for. But that paper didn't appear by chance. A great scientist's second-best paper, or her third, or, for that matter, *many of* her other papers are unusually highly cited as well. Which means that there must be something unique about a researcher who can consistently produce outstanding papers. That unique characteristic is what Q aims to capture. In other words, while luck is important, by itself it won't get you far. The Q-factor helps to turn luck into a consistently high-impact career.

6.3 What Is your *Q*?

The Q-model not only helps us break down the elements associated with high-impact careers, but it also allows us to calculate the Q-factor for each scientist, based on their sequence of publications. The precise solution for Q is somewhat mathematically involved, but once a scientist has published a sufficient number of papers, we can approximate her Q by a rather simple formula [116]. Consider a career in which each paper j published by scientist i collects $c_{10,ij}$ citations in ten years. Start by calculating $\log c_{10,ij}$, for each paper, then averaging over all logarithmic citations. The Q_i is the exponential of that average,

$$Q_i = e^{\langle \log c_{10,i} \rangle - \mu_p}, \tag{6.3}$$

where μ_p is a normalization factor, that depends on the career output of all scientists. Given its dependence on the average of log $c_{10,ij}$, Q is not dominated by a single high- (or low-) impact discovery, but captures, instead, a scientist's *sustained* ability to systematically turn her projects into high (or low) impact publications. To better understand Q, let's look at an example.

The three scientists shown in Fig. 6.2 have comparable productivity, all having published $N \simeq 100$ papers. Yet, their careers have noticeably different impact. Using (6.2), we can calculate the Q-factor for each of them, obtaining Q = 9.99, 3.31, and 1.49, respectively. As Fig. 6.2 illustrates, the Q-factor captures persistent differences in impact across a scientist's sequence of publications: the Q = 9.99 researcher produces one high-impact paper after another. In contrast, the work by the Q = 1.49 scientist garners consistently limited impact. The one in the middle may occasionally publish a

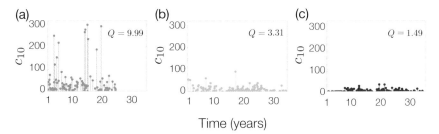

Figure 6.2 Careers with different Q-factor. The figures illustrate career histories of three scientists with comparable productivity ($N \simeq 100$), documenting the notable differences between the papers published by them, given their different Q-factors.

paper that jumps out of the pack – that is, if he gets lucky – but its impact is dwarfed by what the researcher to his left has managed to achieve. Therefore, Q describes scientists' different ability to take random projects r and systematically turn them into high- (or low-) impact publications. Any single one of the projects may be influenced by luck. But inspecting project after project, the true Q of a scientist starts to stand out.

There are many benefits of the mathematical Eq. (6.3). To begin with, it allows us to estimate the expected impact of a career. For example, how many papers does a scientist need to publish before he can expect one of them to reach a certain level of impact? According to (6.3), a scientist with the somewhat low Q = 1.2, similar to that shown in Fig. 6.2c, needs to write at least 100 papers if he wants one of them to acquire 30 citations over a 10-year period. On the other hand, an equally productive scientist with Q = 10 can expect to have at least one of her papers reach 250 citations over the same 10-year period.

Consider next what happens if both scientists ramp up their productivity. Regardless of Q, a higher productivity increases the chance of stumbling upon a fabulous idea, i.e., picking a higher r value. Hence, we expect the highest impact paper to increase for both careers. If the low-Q scientist doubles his productivity though, he'll only enhance the impact of his best paper by seven citations. Compare that with the high-Q scientist, who will see a boost of more than 50. In other words, for a scientist with limited Q, an increase in productivity doesn't substantially improve his chances for a breakthrough – simply working harder isn't enough.

Does the Q-factor increase with age and experience? As scientists, we would like to think that as we advance in our careers, we become better at translating our ideas into high-impact publications. To test the stability of the Q parameter throughout the overall career, we consider careers with at least 50 papers and calculate their early and late Q parameters (Q_{early} and Q_{late}, respectively) using Eq. (6.3) on the first and second half of their papers, respectively. As Fig. 6.3 shows, Q_{late} is proportional to Q_{early}, indicating that the Q parameter does not systematically increase or decrease over a career. In other words, a scientist's Q-factor appears relatively stable during her career. This raises a tantalizing question: can the Q parameter predict the career impact of a scientist?

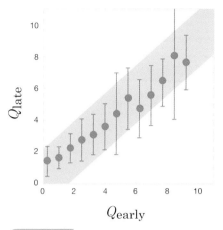

Figure 6.3 The Q-factor appears relatively stable during a career. We compare the Q parameter at early-career (Q_{early}) and late-career (Q_{late}) stage of 823 scientists with at least 50 papers. We measured the two values of the Q parameters using only the first and second half of published papers, respectively. We perform these measurements on the real data (circles) and on randomized careers, where the order of papers is shuffled (shaded areas). For most of the careers (95.1%), the changes between early- and late-career stages fall within the fluctuations predicted by the shuffled careers, suggesting that the Q parameter is relatively stable throughout a career.

6.4 Predicting Impact

To understand if the Q-factor is more effective at gauging the overall impact of an individual's work, we throw several metrics we've discussed so far in this book into a kind of horse race. We start by checking how well these different measures can forecast Nobel prize-winners [116]. For that we can rank physicists based on their productivity N, total number of citations C, citations of the highest impact paper c_{10}^*, h-index, and Q. To compare the performance of each ranking, we use a receiver operating characteristic (ROC) plot that measures the fraction of Nobel laureates at the top of the ranked list. Figure 6.4 shows that, overall, cumulative citation measures, like the number of citations collected by the highest impact paper of a scientist, and the total career citations, do reasonably well. And the h-index is indeed more effective than these citation measures, ranking Nobel-winning careers at a higher accuracy. The worst predictor, interestingly, is the productivity, representing the number of papers published by

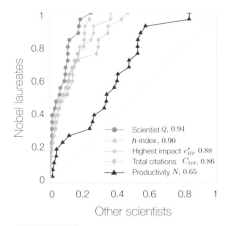

Figure 6.4 Predicting Nobel laureates. ROC plot captures the ranking of scientists based on Q, productivity N, total number of citations C, citations of the highest impact paper c_{10}^* and h-index. Each curve represents the fraction of Nobel laureates versus the fraction of other scientists for a given rank threshold. The diagonal (no-discrimination line) corresponds to random ranking; the area under each curve measures the accuracy to rank Nobel laureates (reported in the legend, with 1 being the maximum). After Sinatra et al. [116].

scientists. In other words, simply publishing a lot is not a path toward the Nobel.

Yet, despite the strength of the h-index, and other citation-based measures, the career-based Q-factor wins out, predicting Nobel careers more accurately than all other measures tested in Fig. 6.4. Therefore, while the h-index remains an excellent measure for the overall impact of a scientist, the Q-factor seems to offer even better predictive power. What does Q capture that h misses?

To understand the difference between Q and h, let's look at a randomized field experiment exploring how economists assess each other's CVs [125]. Professors from 44 universities were randomly selected from among the top 10 percent research-based universities in the world. These professors were asked to rate fake CVs based on the publication histories outlined in those CVs. The fake CVs contained either a publication history that listed only papers that had appeared in elite journals, or a publication history with *both* the elite papers and additional publications that had appeared in lower-tier journals. Economics is an excellent discipline for such a test, because there is a rather stable hierarchy of prestige among the journals. In other words, every economist tends to agree which journals are higher- and which are lower-tier. So, would including publications from well-known, respected, but lower-ranked journals alongside papers from higher-ranked ones make a CV stronger? If so, how much stronger?

The survey respondents were asked to rate the CVs on a scale of 1 to 10, with 1 being the absolute worst and 10 the absolute best. On average, the short CVs listing only top publications received a rating of 8.1. The long CVs – which, remember, contained the *same* top-tier papers as the short CVs but with additional lower-tiered publications – received an average rating of 7.6. This means, a short CV was systematically preferred to a long CV, even though they both featured identical elite publication histories. In other words, the additional publications from lower-tier journals not only didn't help, they actually negatively affected the experts' assessment.

These results are both intuitive and puzzling. On the one hand, they lend evidence to the often-held conjecture that publishing too frequently in lower-ranked journals may work against you. On the other hand, though, these results simply don't make sense if we think in terms of the metrics we use to assess scientific achievement. Indeed, of all the measures we've discussed, from productivity to total number of citations to h-index, each increases monotonically with the number of papers. From this perspective, it will always make sense to publish another paper, even if its impact is limited. First, it will certainly increase your publication count. Second, even if the paper wasn't published in a prestigious journal, with time it may accumulate citations that will contribute to your overall impact, and perhaps even to your h-index. At the very least, it can't hurt.

Except these results show us that it *does* hurt. And that makes sense if we think in terms of the Q-factor. In contrast to other measures, Q doesn't simply grow with an expanding publication list. Instead, it depends on if the additional papers are, on average, better or worse than your other papers. In other words, the purpose of Q is to quantify a scientist's consistent ability to put out high-impact papers over the course of her career. It takes into account all papers, not just the high-impact ones. Hence, if you already have a stellar publication record and publish a few more papers, these additional papers will enhance each of your other metrics, but they aren't guaranteed to increase your Q-factor. In fact, unless your new papers are on par with the papers you typically publish, they may actually lower it.

The superior accuracy of the Q-factor in predicting career impact therefore illustrates that for a career, consistency matters. This conclusion, together with the stability of the Q parameter suggested by Fig. 6.3, paints a picture of rather monotonic careers: we all start our careers with a given Q, high or low. That Q-factor governs the impact of each paper we publish and stays with us until retirement. But is it true? Can we ever break this robotic monotonicity? In other words, do we ever experience periods in our career where we are actually really good at what we do?

7 HOT STREAKS

In physics, 1905 is known as the "miracle year." It is the year that Einstein published four discoveries that changed the discipline forever. By the summer, he'd explained Brownian motion, which pointed to the existence of atoms and molecules; discovered the photoelectric effect, for which he was awarded the Nobel Prize 15 years later, representing a pivotal step toward quantum mechanics; and developed the theory of special relativity, which changed the way we think about time and space altogether. Then, before the year ended, he scribbled down the world's most famous equation: $E = mc^2$.

Einstein's 1905 can be described as a "hot streak," a burst of outstanding achievements. Hot streak is a familiar term in sports or gambling. If in basketball you land two shots without even touching the rim, you are not at all surprised when you land that third one. You probably could even "feel" that strange and magical momentum. While for decades hot streaks were taken as a given, in 1985 a much-quoted psychology study contested their existence [126], concluding that basketball is no streakier than a coin toss. The finding suggested that the concept is a fallacy, rooted in our psychological biases for small numbers – it only *feels* that you're on a roll, yet in reality that streak of luck is what one would expect by chance. The 1985 study triggered a still-ongoing debate among psychologists and statisticians [127–130], with the latest statistics papers arguing that the fallacy may itself be a fallacy, and hot streaks do seem to exist in sports after all [127, 131]. Yet, as the debate unfolds in sports, gambling, and financial markets, it raises an intriguing question: If we define our career as the sequence of papers a scientist produces over a lifetime, do we ever experience hot streaks in our career?

7.1   Bursts of Hits

Across the careers of scientists, artists, and film directors, we've seen that the three biggest hits of a career each occur at random within the sequence of works one produces. This finding tells us that creativity is random and unpredictable, with chance playing an outsized role in the timing of key achievements. In other words, scientific careers are unlikely to include hot streaks. Yet, at the same time, the random impact rule raises a puzzling question: What happens after we finally produce a breakthrough?

The Matthew effect tells us that winning begets more winnings. So, if we produce one big hit, even if its timing is random, it would follow that we'd produce more hits afterwards. Yet according to the random impact rule, the opposite seems to be true. Indeed, if the impact of each work in a career is truly random, then one's next work after a hit may be more mediocre than spectacular, reflecting regression toward the mean. So, are we really regressing toward mediocrity after we break through?

To answer this question, we examined the *relative* timing of hit works within a career [117]. Specifically, we asked: Given when someone produced their best work, when would their second-best work be? To measure the correlation between the timing of the two biggest hits within a career (e.g., N^* and N^{**}) we calculate the joint probability $P(N^*, N^{**})$ for the two of them to occur together, and compare it with a null hypothesis in which N^* and N^{**} each occur at random on their own. Mathematically, this is captured by the normalized joint probability, $\varphi(N^*, N^{**}) = P(N^*, N^{**})/P(N^*)P(N^{**})$, which is best represented as a matrix (Fig. 7.1). If $\varphi(N^*, N^{**})$ is approximately 1, then the chance to find the biggest and the second biggest hit to occur together within a career is about what we would expect if both occur at random. If, however, $\varphi(N^*, N^{**})$ is above 1, it means that N^* and N^{**} are more likely to appear together, corresponding to a correlation that is not anticipated by the random impact rule.

Figure 7.1 shows $\varphi(N^*, N^{**})$ for careers across sciences, movies, and arts, leading to three important conclusions:

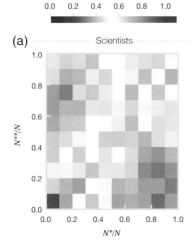
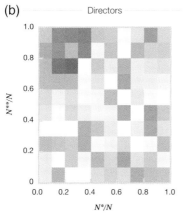
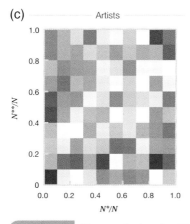

Figure 7.1 Hot-streaks in scientific, cultural, and artistic careers. $\varphi(N^*, N^{**})$, color-coded, measures the joint probability of the top two highest-impact works within the career of (a) scientists, (b) film directors, and (c) artists. $\varphi(N^*, N^{**}) > 1$ (red along diagonal) indicates that two hits are more likely to occur next to each other than expected by chance. After Liu et al. [117].

- First, $\varphi(N^*, N^{**})$ is notably higher along the diagonal elements of matrices, indicating that N^* and N^{**} are much more likely to co-locate with each other than expected by chance. In other words, if we know where your biggest hit happens in the course of your career, we would know quite well where your second biggest hit is – it will be just around the corner of your biggest hit. The two most important works of a scientist are on average 1.57 times more likely to occur back-to-back than we'd expect to see by chance.
- Second, φ features a roughly even split across the diagonal, which means that there is a comparable likelihood of the biggest hit arriving either before or after the second biggest hit.
- Third, the co-location pattern is not limited to the two highest impact works within a career. Indeed, if we repeat our analyses for other pairs of hits, such as N^* vs. N^{***} and N^{**} vs. N^{***}, we find the same pattern. It is not only the top two hits which appear close to each other. The *third* hit is also in the vicinity of the first two.

These results offer a more nuanced portrait of individual careers. Indeed, while the timing of each highest impact works is random, *the relative timing of the top papers in an individual's career* follows highly predictable patterns. As such, individual career trajectories are not truly random. Rather, hit works within a career show a high degree of temporal clustering, with each career being characterized by bursts of high-impact works coming in close succession. Further, these patterns are not limited to scientific careers. The hits of artists and film directors show similar temporal patterns, suggesting that there's a ubiquitous clustering of successful work emerging across a range of creative careers. What are the mechanisms responsible for these remarkable patterns?

7.2 The Hot-Streak Model

To understand the origin of the observed bursts of career hits, let's start with the Q-model discussed in the previous chapter. Imagine that every time a scientist publishes a paper, the work's impact is determined by a random number generated from a given distribution, which is fixed for that individual. Since citations are well approximated by lognormal distributions, let's assume for convenience that their logarithm is drawn from a normal distribution, with average Γ_i.

A career generated by this null model will nicely follow the random impact rule: each hit, including the biggest one, occurs randomly within a career [2, 116]. It all depends on luck – the pay-off of a lottery ticket drawn from a distribution of random numbers.

However, this null model fails to capture the temporal correlations documented in Fig. 7.1. The main reason is shown in Fig. 7.2a–c, where, for illustration purposes, we selected one individual from each of the three domains and measured the dynamics of Γ_i during his or her career. These examples indicate that Γ_i is not constant. Rather, it is characterized by a baseline performance (Γ_o) until a certain point in a career, after which it is elevated to a higher value Γ_H ($\Gamma_H > \Gamma_o$). That elevated performance is sustained for some time before eventually falling back to a level similar to Γ_o. This observation raises an interesting question: Could a simple model, which assumes everyone experiences a short period like Γ_H, explain the patterns documented in Fig. 7.1?

Remember, this new model introduces only a simple variation over the Q-model – a brief period of elevated impact. But interestingly, that small change can account for the empirical observations that the random impact rule or the Q model failed to capture (Fig. 7.1). During the period in which Γ_H operates, the individual seemingly performs at a higher level than her typical performance (Γ_o). We call this model the *hot-streak model*, where the Γ_H period corresponds to the hot streak.

The true value of the hot-streak model emerges when we apply it to career data, allowing us to obtain the parameters which characterize the timing and the magnitude of hot streaks for individual scientists, leading to several important observations:

(1) The measurements indicate that hot streaks are ubiquitous across creative careers. About 90 percent of scientists experience at least one hot streak, as do 91 percent of artists and 82 percent of film directors. This means that hot streaks are not limited to Einstein's miracle year of 1905. The director Peter Jackson's hot streak lasted for 3 years, while he directed *The Lord of the Rings* series, and Vincent van Gogh's occurred in 1888, the year he moved from Paris to Southern France, producing works like *The Yellow House, Van Gogh's Chair, Bedroom in Arles, The Night Café, Starry Night Over the Rhone*, and *Still Life: Vase with Twelve Sunflowers*.

(2) Despite their ubiquity, hot streaks are usually unique in a creative career. Indeed, when we relaxed our algorithm to allow for up to

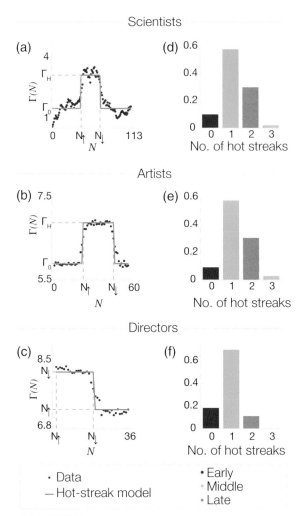

Figure 7.2 **The hot-streak model.** (a–c) $\Gamma(N)$ for one scientist (a), artist (b) and film director (c), selected for illustration purposes. It shows that real careers can have strong temporal variations, characterized by brief period of elevated impact. (d–f) Histogram of the number of hot streaks in a career shows that hot streaks are ubiquitous yet usually unique. Most people have hot streaks, and they mostly likely have just one. After Liu et al. [117].

three hot streaks, we found that 68 percent of high-impact scientists experience only one hot streak (Fig. 7.2d). A second hot streak may occur but is less likely, and occurrences of more than two are rare.
(3) Hot streaks occur randomly within a career, offering a further explanation for the random impact rule. Indeed, if the hot streak

occurs randomly within the sequence of works produced in a career, and the highest impact works are statistically more likely to appear within the hot streak, then the timing of the highest impact works would also appear to be random.

(4) Hot streaks last for a relatively short period. The duration distribution of hot streaks peaks around 3.7 years for scientists, indicating that, whether it happens early or late in a scientist's career, it most likely lasts for only 4 years. For artists or film directors, it lasts about 5 years.

(5) While people create more impactful work during hot streaks, they aren't more productive than we would expect during that time. It's just that what they do produce is substantially *better* relative to the rest of their work.

Why do we have hot streaks in the first place? There are several plausible hypotheses. For example, innovators may stumble upon a groundbreaking or temporally resonant idea, which then results in multiple high-impact publications. Alternatively, since high-impact outcomes are increasingly produced by teams, an individual's hot streak may reflect

---

Box 7.1 Bursts in human activities

Short periods of elevated performance are not limited to creative achievements: bursts are also observed in a wide range of human activity patterns. For example, it is often assumed that the timing of specific human activity, like making phone calls, is random. If it is truly random, the inter-event times between consecutive events should follow an exponential distribution. The measurements indicate otherwise: The inter-event times in most human activities is well approximated by a power law distribution [132–134]. This means that the sequence of events is characterized by bursts of activities, occurring within a relatively short time frame, and occasionally there are long time gaps between two events. The "bursty" pattern of human behavior has been documented in a number of activities, from email communications to call patterns and sexual contacts. Although the burstiness and hot streaks are measured differently, these examples suggest that hot streaks have commonalities with the bursts observed in human activities, and the hot-streak phenomena may have relevance beyond human creativity.

a fruitful, repeated, but short-lasting collaboration. Or perhaps it is related to shifting institutional factors like tenure track, corresponding to career opportunities that augment impact but last for a fixed duration. Analyses of real careers suggest that, while plausible, none of these hypotheses alone can explain the observed hot-streak phenomena. Instead, our preliminary analyses of scientific careers suggest that a particular combination of research strategies, namely exploration followed by exploitation, seems to be particularly powerful in predicting the beginning of a hot streak. As more data on individual careers becomes available, we may be able to identify the drivers and triggers for hot streaks, helping us answer a range of new questions. Can we anticipate the start and the end of a hot streak? Can we create an environment to facilitate and promote the onset of a hot streak, and to extend it when it emerges?

7.3 What Do Hot Streaks Mean for Us?

In science, future impact is critical for hiring, career advancement, research support, and other decisions. Yet, the hot-streak phenomenon shows that performance can be rather uneven: scientists have substantially higher performance during a hot streak. Indeed, the timing and magnitude of hot streaks dominate a researcher's career impact, measured by the total citations collected by all of her papers. As such, ignoring the existence of hot streaks – and their singular nature – may lead us to systematically over- or under-estimate the future impact of a career.

For example, if a hot streak comes early in a career, it would lead to a period of high impact that peaks early. That impact may diminish, however, unless a second hot streak emerges. On the other hand, if an individual has not yet experienced her hot streak, judging her career based on her current impact may underestimate her future potential. The hot streak phenomenon is also quite relevant to funding agencies, given that hot streaks and research grant both last about four years, raising the question of how can funding best maximize its impact on individual careers.

But, admittedly, implementing changes in light of the hot streak phenomenon is a challenge. It would be absurd for a junior scientist to justify his middling dossier to his tenure committee by saying, "My hot streak is coming!" Similar puzzles confront prize committees when the prize has an age threshold. The Fields Medal, the highest reward in mathematics, only recognizes mathematicians under 40. The National Science Fund for Distinguished Young Scholars, an important milestone of a successful scientific career in China, excludes applicants over 45. Given these arbitrary age thresholds and the randomness of hot streaks, it means that a substantial portion of well-deserving candidates will miss out if they experience their big break late.

Further, our preliminary analysis also suggests that incorporating hot streaks in funding decision may not be as intuitive as it appears either. For example, we examined the relationship between the timing of NIH awards and the beginning of hot streaks for their PIs, finding that PIs are more likely to have initiated a hot streak than expected *prior to* being awarded their first R01 grant. In other words, a scientist's hot streak did not follow funding. Rather, it was funding that follows hot streaks. These findings are consistent with NIH's emphases on preliminary results, and they also appear reasonable from a funder's perspective, as people who have had a hot streak tend to have higher impact (hence worthy of funding). Yet at the same time, these findings raise the question of whether – by associating funding decisions with past success – funders may miss the critical period when PIs are most creative, especially considering the fact that hot streaks are usually unique in an individual career.

Taken together, if our goal is to identify and nurture individuals who are more likely to have a lasting impact in their field, we must consider incorporating the notion of hot streaks into the calculus. If we don't, we may miss out on vitally important contributions. Yale University learned this the hard way.

We already encountered John Fenn, and his late discovery. It was at age 67, when he published a study that identified a novel electrospray ion source. It was a real breakthrough – or at least he thought so. But Yale still pushed him out the door, by sending him into retirement. After he reluctantly relocated to Virginia Commonwealth University, which provided him with the lab he needed to continue his investigations, he embarked on a classical hot streak. Between 1984 and 1989, he published one study after another, ultimately developing electrospray ionization for mass spectrometry, which enabled faster, more accurate measurement of large molecules and proteins, spurring multiple innovations in cancer diagnosis and therapy. That five-year hot streak, which took place during Fenn's forced retirement, ultimately defined his career, and earned him the 2002 Nobel Prize in chemistry. His hot streak is so evident that you can see directly just by inspecting Fig. 5.5. When is Fenn's most-cited paper? And when is his second highest? What about the third? They all fall within the same five-year window.

In Fenn's career, we see perhaps the most important – and uplifting – implication of the hot-streak phenomenon for the many individual scientists out there striving to make their mark on the world. Remember that the conventional view, which we discussed in depth in Chapter 3, is this: Our best work will likely happen in our 30s or 40s, when we have a solid base of experience and the energy and enthusiasm to sustain high productivity; once we pass the mid-career point, our hopes of a breakthrough start to dim. The hot streak phenomenon, coupled with the random impact rule contradicts this, indicating, instead, that a hot streak can emerge at any stage of your career, resulting in a cluster of high-impact works.

Therefore, this chapter offers hope: Each new gray hair, literal or figurative, does not by itself make us obsolete. As long as we keep putting work out into the world like Fenn – drawing cards from the deck in search of a lucky ace – our hot streak could be just around the corner.

This means that while we may not all have Einstein or Fenn's impact, if we keep trying, our own version of a miracle year may still be ahead, just out of sight.

Part II: THE SCIENCE OF COLLABORATION

In 2015, Marco Drago was a postdoctoral researcher at the Albert Einstein Institute in Hannover, Germany, working on the Laser Interferometer Gravitational-Wave Observatory (LIGO) experiment. The project's goal was to detect gravitational waves – ripples in the fabric of spacetime caused by the collision and merging of two black holes. Drago spent most of his time overseeing one of four data "pipelines" – automated computer systems which scoured the raw data coming from the LIGO detectors for unusual signals. Just before lunch on September 14, 2015, while on the phone with a colleague, he received an automated email alert notifying him that the 4 km-long tunnels housing sensitive laser beams in ultrahigh vacuum, had just vibrated. Such an event was not uncommon, but as Drago clicked open the email, he quickly realized this one was different. The event it reported was *huge* by LIGO's standards. While the vibration was less than a trillionth of an inch, that was more than double the size of a typical event. It was so large, in fact, that Drago instantly dismissed it. It seemed too good to be true.

It took another five months to officially confirm the truth behind the September 14 event. As announced on February 12, 2016, the event registered by the automated email had been the real deal. That tiny signal told us that about 1.3 billion years ago, across the Universe, two black holes collided, forming a new black hole that was 62 times as heavy as our Sun. Their collision radiated 100 times more energy than all the stars in the Universe combined, generating gravitational waves that rippled in every direction. Weakening as they traveled through space at the speed of light, the waves finally reached Earth on that quiet September morning in 2015, shaking the detectors by about a thousandth of the diameter of a proton. That vibration validated the prediction of Albert Einstein's general theory of relativity, becoming, according to some, the "discovery of the twenty-first century."

While Drago was the lucky person who saw the signal first, it wasn't really *his* discovery. An international team of scientists spent over 40 years building the experiment, and when the paper reporting the detection of gravitational waves was finally published, it listed over 1,000 authors from all over the world [135]. There were the physicists who dreamed up the project and calculated its feasibility, the engineers who designed the tunnels, and the administrative staff who oversaw the day to day operations. At US$1.1 billion, LIGO is still the largest and most ambitious project ever funded by the US National Science Foundation (NSF).

By contrast, when the Prussian Academy of Science first heard about what would be deemed the "discovery of the twentieth century" in November 1915, it had a single author, Albert Einstein. Occurring exactly 100 years apart, these two discoveries reveal an important way that science has changed over the past century. We imagine science as a solitary endeavor, picturing Einstein, Darwin, and Hawking on solo journeys, inching toward that "Eureka!" moment. Yet today, most science is carried out by teams [136, 137]. Indeed, 90 percent of all science and engineering publications are written by multiple authors. These can be two-person teams, such as Watson and Crick, who unveiled the structure of DNA, but many important discoveries are made through large-scale collaborations, like those pursued at CERN, the Manhattan Project, or Project Apollo. Scientific teams have produced breakthroughs that could not have been achieved by lone investigators. And such large-scale team projects often have an enormous impact, not just on science, but on the economy and society. Take for example the Human Genome Project. It not only jump started the genomic revolution; its direct expenditures of about US$3.8 billion also generated US$796 billion in economic growth and created 310,000 jobs [138]. However, these kinds of massive collaborations also introduce a new set of unique challenges for scientists, from team communication to coordination, which, if left unaddressed, could jeopardize the success of their projects.

Why are some collaborations impactful, fruitful, and lasting, while others fail (at times, spectacularly)? What factors help or hinder the effectiveness of teams? How do we assemble a highly productive team? Is there an optimal size for the team? How do teams evolve and dissolve over time, and how can they maintain and diversify their membership? And how do we assign credit for a team's work? In Part II, we focus on the rich body of literature that is often called the science of team science (SciTS), examining how scientists collaborate and work together in teams. We start by asking a simple question: Do teams matter in science?

8 THE INCREASING DOMINANCE OF TEAMS IN SCIENCE

To what degree do teams contribute to the production of science in the twenty-first century? The answer is provided by a study that explored the authorship of 19.9 million research articles and 2.1 million patents [136], revealing a nearly universal shift towards teams in all branches of science (Fig. 8.1a). For example, in 1955, nearly half of all science and engineering publications were by single authors, but by 2000, the number of solo-authored papers had dwindled dramatically, while team-authored papers now made up 80 percent of all publications. Importantly, the shift toward teams is not simply driven by the fact that the experimental challenges are becoming larger, more complex, and more expensive. Pencil-and-paper disciplines like mathematics and the social sciences exhibit the same patterns. Teams wrote only 17.5 percent of social science papers in 1955, but, by 2000, team-based papers became the majority, reaching 51.5 percent – witnessing the same trend toward teamwork as had been observed decades earlier in the natural sciences.

But perhaps more interesting than the trend itself is the kind of research that team work has produced. Teams do not merely produce more science; they are increasingly responsible for discoveries with larger impacts [136]. Indeed, on average, team-authored papers garner more citations than single-authored work at all points in time and across all broad research areas (Fig. 8.1b).[1]

---

[1] It may be tempting to attribute the higher impact of team-authored papers to self-promotion: Authors like to cite their own work, so the more authors a paper has, the more citations it's guaranteed to get. Yet, the higher impact of team-authored paper remains unchanged if we remove self-citations [139, 140].

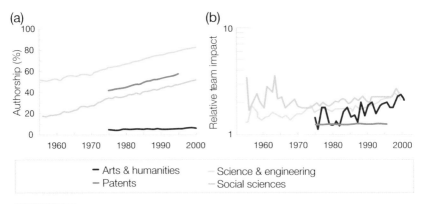

Figure 8.1 The growing dominance of teams. (a) Changes in the fraction of papers and patents written by teams over the past five decades. Each line represents the arithmetic average taken over all subfields in each year, with colors indicating different fields. (b) The relative team impact (RTI) represents the mean number of citations received by team-authored work divided by the mean number of citations received by solo-authored work in the same field. A ratio of 1 implies that team- and solo-authored work have comparable impact. The lines present the arithmetic average of RTI in a given year for the entire field. After Wuchty et al. [136].

The influence of teams is especially interesting if we focus exclusively on the very top tier of high-impact papers: Teams are increasingly responsible for producing the most important scientific hits. Indeed, whereas in the early 1950s, the most cited paper in a field was more likely to have been written by a lone author than a team, that pattern reversed decades ago [139]. Today in science and engineering a team-authored paper is 6.3 times more likely than a solo-authored paper to receive at least 1,000 citations. This pattern is even more pronounced in the arts, humanities, and patents, where since 1950s teams have *always* been more likely than individuals to produce the most influential work.

8.1 The Rise of Team Science

Why do teams dominate science to such a degree? One hypothesis is that teams excel in generating truly novel ideas. By grouping experts with diverse but specialized knowledge, teams give their members access to a broader pool of knowledge than any individual collaborator could have. By harnessing varied methodologies and bodies of research, teams are more capable of creating innovative combinations of ideas and concepts. Indeed, as we will learn in later chapters (Chapter 17), papers that introduce novel combinations while also remaining embedded in conventional thinking increase their chance of becoming hits by at least twofold [92] – and teams are 37.7 percent more likely than solo authors to insert novel combinations into familiar knowledge domains [92]. Since the greatest challenges of modern science often require interdisciplinary expertise, teams are emerging as the key innovation engine to produce tomorrow's breakthroughs.

For the individual researcher, there are also other benefits to working in teams [141]. For example, colleagues working together can bounce ideas off one another and check one another's work, aiding both innovation and rigor. Collaboration also boosts the visibility of a researcher by exposing her publications to new coauthors and disciplines, resulting in a broader and more diverse audience. Additionally, teamwork not only leads to a wider readership, but also helps in

Box 8.1 Ageless teamwork

Collaboration in the creative space is not a modern invention. The *Yongle Encyclopedia*, or *Yongle dadian* ("The Great Canon of the Yongle Era"), is the world's largest paper-based encyclopedia, comprising 22,937 rolls or chapters in 11,095 volumes. Commissioned by the Yongle Emperor of the Ming dynasty in 1403 in China, it required the collaborative effort of 2,000 scholars over a period of five years.

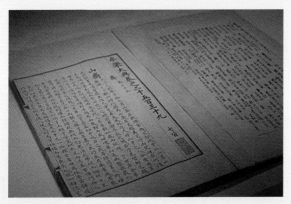

Figure 8.2 The *Yongle Encyclopedia*, on display at The National Library of China in 2014.

securing funding for further research. Indeed, a study focusing on a sample of 2,034 faculty members at Stanford University over a 15-year period found that collaborations produce more grant submissions, are more likely to be awarded grants, and yield larger dollar amounts [142].

It is therefore not hard to imagine that the best and brightest scientists are those who are more willing to collaborate. In 1963, the sociologist Harriet Zuckerman set out to discover how outstanding scientists worked, interviewing 41 out of the 55 Nobel laureates living in the US at that time. She found that an internal bent toward teamwork seemed to be a trait that they all shared. The Nobel laureates were more willing to collaborate than their less prominent colleagues, which offered them a clear, long-term advantage in science [101].

8.2   The Drivers of Team Science

The recent explosion of teamwork can be attributed to two major trends. First, as science grows increasingly complex, the instruments required to expand the frontier of knowledge have increased in scale and precision. For example, the Large Hadron Collider (LHC), the world's largest particle collider at CERN, is crucial for advances in particle physics. However, such a tool would be unaffordable to individual investigators or institutions. The LHC – both its conception and its financing – would have been unthinkable without teamwork. Over 10,000 scientists and engineers from over 100 countries contributed to the project. Hence collaboration is not just beneficial, it's *necessary*, forcing the community to pool resources together to help advance scientific understanding.

Second, the ever-broadening body of human knowledge is so vast that it is impossible for any one person to know everything. Even for those working in relatively small, esoteric fields, the burden of knowledge on successive generations of scientists is continually increasing. To cope, scientists often specialize, narrowing their focus in order to manage the knowledge demand and to reach the frontiers of knowledge faster. Such specialization has led to the "death of the renaissance man" in science, a phenomenon documented in the context of inventions [104]. Indeed, data on consecutive patent applications filed by solo inventors show that individuals increasingly remain within a single technological area and demonstrate a decreased capacity to contribute new, unrelated inventions. Thus, as the burden of knowledge required to move between fields becomes more overwhelming, collaborations become one way to reach beyond one's own specialized subfield [104, 105]. In other words, scientists not only want to collaborate – they *have to*. Increasing specialization means that each person has an excellent command of one piece of the puzzle. But to address the complex problems faced by modern science, scientists need to bring these pieces together, melding diverse skill-sets and knowledge in ways that allow us to innovate.

8.3 The Death of Distance

With worldwide Internet access and increasingly inexpensive transportation, it is now easier than ever to collaborate across borders, overcoming traditional geographical constraints (Fig. 8.3). As the figure shows, between 1900 and 1924, collaborations across different institutions were prominent only in the US; international collaborations were mainly between the US and the UK. However, both types of collaborations were relatively rare. Between 1925 and 1949, international collaborations started to form between India and the UK, as well as between

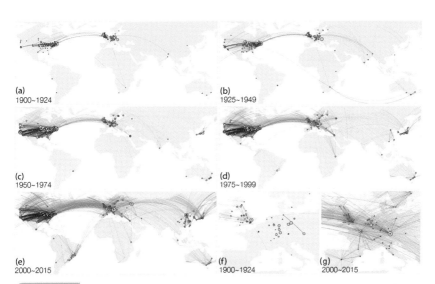

Figure 8.3 **A brief history of globalization of science collaborations.** The figure shows two types of collaborations: across institutions within the same country (blue/green) and across institutions of different countries (red). Circles represent the top 200 most cited institutions in the world. Each circle's size is proportional to the institution's citation count. After Dong et al. [4].

Australia and the US. Due to World War II collaborations in Europe shrank during this period. Meanwhile, collaborative relationships in America were rapidly developing in the Midwest. Between 1950 and 1974, Israel and Japan started to engage in international teamwork. At the same time, the West Coast and the Southern United States became hubs of scientific collaboration. Between 1975 and 1999, Africa began to develop relationships with Europe. Surprisingly, within-institution collaborations in the US decreased relative to those in Europe, although the absolute number of collaborations grew substantially over time for all countries. In the twenty-first century, more and more countries have risen to the international stage to collaborate extensively with others. Indeed, teams increasingly span institutional and national boundaries. Analyzing 4.2 million papers published between 1975 and 2005, researchers distinguished three types of authorship: solo authors, collaborators working at the same university, and collaborators who bridged the gap between institutions [143]. Of the three, collaborations between university faculty at different institutions was the fastest – and, in fact, the only steadily growing authorship structure. During the 30-year period examined, such inter-institutional collaborations quadrupled in science and engineering, reaching 32.8 percent of all papers published (Fig. 8.4a). The social sciences experienced similar trends, with their share of papers written in multi-university collaborations rising even more rapidly over the period and peaking at 34.4 percent (Fig. 8.4b).

Teams today are increasingly spanning national boundaries as well. Worldwide, between 1988 and 2005, the share of publications with authors from multiple countries increased from 8 percent to 20 percent [144]. Another analysis of papers published between 1981 and 2012 calculated the balance of international and domestic research collaboration in different countries [145]. As shown in Fig. 8.5, while total research output has grown substantially over time, domestic output has flat-lined in the United States and in western European countries. This means that these days, international collaborations fuel the growth of science in these countries. By contrast, in emerging economies, international collaborations have yet to match – much less eclipse – domestic output. The total volume of papers from China, Brazil, India, and South Korea has increased 20-fold, from fewer than 15,000 papers annually in 1981 to more than 300,000 papers in 2011. Yet, about 75 percent of the research output from these four countries remains entirely domestic (right column in Fig. 8.5).

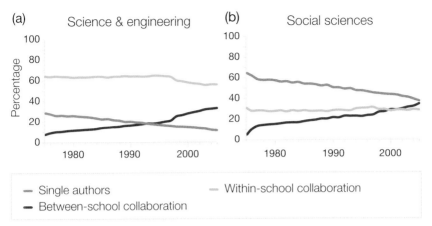

Figure 8.4 The rise in multi-university collaboration. By comparing the percentage of papers produced by different authorship combinations, the plots document the increasing share of multi-university collaborations between 1975 and 2005. This rise is especially strong in science and engineering (a) and social sciences (b), whereas it remains weak in arts and humanities, in which collaboration of any kind is rare [143]. The share of single-university collaborations remains roughly constant with time, whereas the share of solo-authored papers strongly declined in science & engineering and social sciences. After Jones et al. [143].

The growth we see in international collaborations bodes well for a few reasons. We already know that citation impact is typically greater for teams than for solo authors. That benefit strengthens when the collaborators in question are either international [145] or between universities [143]. In both the UK and US, papers with shared international authorship are cited more than papers with only domestic authors. And the benefit appears to be growing – between 2001 and 2011, this "impact premium" rose in both countries by 20 percent. Multi-university collaborations have a similar impact advantage [143]: When an entire team comes from the same university, their probability of publishing a paper with above-average citations is around 32.7 percent in science and engineering and 34.1 percent in the social sciences. But, if the collaboration involves different institutions, that probability increases by 2.9 percent or 5.8 percent, respectively.

While the advantages of collaborations across geographies are clear, an important consequence of the "death of distance" in modern science is increasing inequality in both impact and access to resources [12, 141, 146]. Indeed, scientists often reach across university boundaries as they select coauthors, but rarely across university prestige levels.

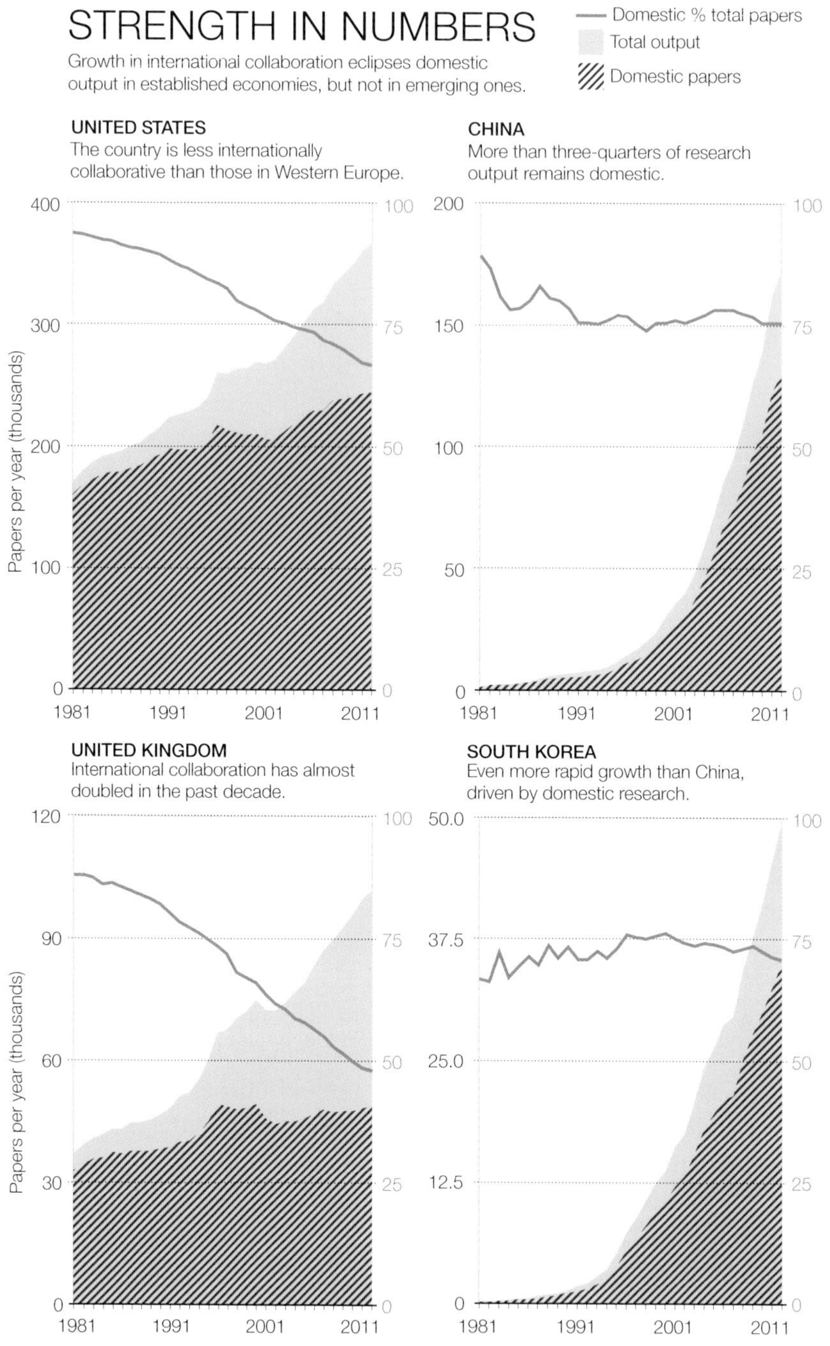

STRENGTH IN NUMBERS

Growth in international collaboration eclipses domestic output in established economies, but not in emerging ones.

— Domestic % total papers
▒ Total output
▨ Domestic papers

UNITED STATES
The country is less internationally collaborative than those in Western Europe.

CHINA
More than three-quarters of research output remains domestic.

UNITED KINGDOM
International collaboration has almost doubled in the past decade.

SOUTH KOREA
Even more rapid growth than China, driven by domestic research.

Figure 8.5 **The increasing role of international collaboration.** If a paper only contains authors whose addresses are from the home country, then it is counted as domestic output of that country. Comparing the left and right panels shows that growth in international collaboration eclipses the growth of domestic output in established economies, but not in emerging ones. After Adams [145].

Researchers at elite universities are more likely to collaborate with scientists at other elite universities, and scientists at lower-ranked institutions tend to collaborate with researchers from institutions of comparable rank [143]. Hence, even as geographic distance loses its importance, the role of social distance is increasing. This kind of stratification among institutions may further intensify inequality for individual scientists.

Furthermore, the benefits of collaboration vary greatly based on a scientist's position in a collaboration network. For instance, if an African country devotes special funding to help its researchers collaborate with their US-based colleagues, which American institution will the researchers approach first: an Ivy League university, or a small liberal-arts college? Unsurprisingly, they will likely seek out well-known researchers at elite institutions. That means successful scientists at prestigious institutions are more likely to benefit from the resources offered by the global collaborative networks. Given that multi-university publications have a larger impact than collaborations within the same university, and that researchers at elite universities are also more likely to engage in such cross-university collaborations, the production of outstanding science may become more and more concentrated at elite institutions [141, 143].

Although team science intensifies inequality for individual scientists and institutions, it has benefits across the globe [146]. Today, a successful scientist at a prestigious university in America or Europe can design studies to be carried out by collaborators in less-developed countries, such as China, where labor-intensive scientific work is less expensive. Such collaborations can offer mutual benefits and help reduce the gap in knowledge between more- and less-developed countries.

On the other hand, the globalization of science also has dire implications for "brain drain" [145], which may fuel a growing divide between international and domestic research. Indeed, as science becomes more international, every nation will be able to more easily connect to the same global knowledge base, making it possible for their brightest scientists to continue their research elsewhere. This, in turn, compromises human resources in science for individual nations. Therefore, understanding the nuanced dynamics of team science may be necessary for nations to stay competitive, helping them to retain and attract global talents amid the growing scarcity of truly able researchers.

*

As we showed in this chapter, teams play an increasingly important role in driving the overall production of knowledge, as well as key breakthroughs. Once a rarity, team science has become a dominant force in science, a phenomenon that cannot be ignored by scientists, policymakers, or funding agencies. Yet, despite the many benefits teamwork affords, there are also reasons to believe that teams are not optimized for discovery [137, 141, 147]. For example, while collaboration lends legitimacy to an idea – after all, a whole team needs to believe in a project to pursue it [147] – colleagues must reach consensus before they can move forward, a process which can hinder the group's headway. And while collaborations help researchers bridge disciplinary boundaries and generate new insights and hypotheses, the time and energy they must devote to coordination and communication can be costly, in some cases outweighing the benefits garnered by working collaboratively. This tradeoff is further complicated by the increased ambiguity about who should, and does, receive credit for the collaborative work. Indeed, strategically balancing these complex trade-offs is critical to building successful teams in science, and that's what we will discuss in the next chapters.

9 THE INVISIBLE COLLEGE

We tend to believe that rubbing shoulders with the top performers in their field will also make us great. But do our peers truly make us better scientists? In the previous chapter, we have explored how collaboration influences the impact of a project. But how does collaboration influence the productivity and the impact of individual scientists? This question encompasses a broader set of problems, known in the economics literature as "peer effects" [148–151]. To see how powerful peer effects can be, let's visit the dormitories at Dartmouth College [149].

Freshmen entering Dartmouth College are randomly assigned to dorms and to roommates. However, a study found that when students with low grade-point averages (GPA) were assigned to a room with higher-scoring students, their grades improved. Since students with different majors lived together, we can't attribute the improvement to something like weaker students getting help from their roommates on shared coursework. Rather, they became better students *because* they were surrounded by higher-performing peers.

Peer effects also come into play when people work together towards common goals. Indeed, over the past 20 years, numerous studies have documented peer effects in diverse occupations, from fruit pickers, to supermarket cashiers, to physicians, to sales teams, and more [151]. For example, examining a national supermarket chain, researchers found that when a less productive cashier was replaced with a more productive cashier on a shift, other cashiers on the same shift scanned items more quickly [150]. This change in behavior seems counterintuitive – after all, if a star cashier replaces a slower scanner, other cashiers could simply take it easy, engaging in the so-called "free riding" phenomenon. Yet, the cashiers acted in the exact opposite way: The presence of their productive peer nudged them to enhance their performance.

Understanding how we are influenced by others is important, because it could unleash social multiplier effects, eventually improving everyone's outcome. Consider again the college dorm example: If the academic performance of some students improves, then their improved performance could in turn improve the outcomes of others in their respective peer groups and continue rippling out to more and more students.

But do peer effects exist in science? And if so, how can students, scientists, administrators, and policymakers take advantage of them?

> **Box 9.1 Mencius' mother, three moves**
>
> The Chinese philosopher, Mencius (372–289 BCE), is the most famous Confucian after Confucius himself. One of the most famous Chinese four-character idioms is Meng Mu San Qian, meaning "Mencius' mother, three moves" a phrase that captures one of the earliest examples of peer influence. It describes why Mencius' mother moved houses three times before finding a place she felt was suitable for her child's upbringing. Mencius' father died when he was little, so Mencius' mother, Zhang, raised Mencius alone. Poor as they were, their first home was next to a cemetery, where people loudly mourned for the dead. Soon Mencius' mother noticed that her son had started to imitate the paid mourners. Horrified, she decided to move near a market in town. It wasn't long before the young Mencius began to imitate the cries of merchants, a profession of ill-repute in early China. So, Zhang, distraught, decided to move again, this time landing in a house next to a school. There Mencius began to spend time with the students and scholars, imitating their habits and becoming more focused on his own studies. Pleased to see such behavior in her son, and attributing it to his proximity to the scholars, Zhang decided to make the house by the school her permanent home, ensuring that Mencius would become Mencius.

9.1 A Bright Ambiance

Following numerous interviews with Nobel laureates, Robert Merton noted the "bright ambiance" that such star scientists often provide [60]. That is, star scientists not only achieve excellence themselves; they also have the capacity to evoke excellence in others. We may measure this capacity by examining changes in a department after it hires a new star scientist. In one study, researchers traced the publication and citation records of each member of a department after a star scientist arrived, examining 255 evolutionary biology departments and 149,947 articles written by their members between 1980 and 2008 [152]. They found that after a star arrived, the department-level output (measured in number of publications, with each paper weighted by its citation count) increased by 54 percent. This increase could not be attributed to the publications of the stars themselves: After removing the direct contributions of the star, the output of the department *still* increased by 48 percent. In other words, much of the observed gains came from the increased productivity of the colleagues who'd been there before the star arrived. And importantly, the productivity gain persisted over time: Even eight years after the star's arrival, the productivity of her colleagues had not diminished.

It's worth noting that those individuals in the department whose research was the most related to the star – i.e., those who have cited her papers before – saw the biggest boost in productivity. By contrast, colleagues whose research was unrelated to the new hire were far less influenced.

Yet the star's biggest contribution to her new department comes not in productivity increases but in the quality of future hires. Indeed, the quality of subsequent faculty hires (measured by the average citations of papers published by a scientist at the time she joined the department) jumped by 68 percent after the arrival of a star. Differentiating again between the new hires related and unrelated to the star's research, the quality of related new hires increased by a staggering 434 percent. Furthermore, the quality of unrelated new hires also increased by 48 percent. It seems that stars not only shine on their peers, prompting them to succeed, but also attract other future stars to join them.

Box 9.2 Causal inference for peer effects

In observational studies, i.e., those that rely on pre-existing datasets to draw conclusions, peer effects are difficult to quantify, hence insights obtained in such data-driven studies must be interpreted with care. For instance, is it possible that stars do not *cause* the rise in productivity, but rather they are attracted to already improving departments? Or could it be that an unobserved variable, like a positive shock to department resources (e.g., philanthropic gifts, increase in government funding, the construction of a new building) offered the opportunity for the department to hire a star – and that the same shock increased its incumbent productivity and the quality of subsequent recruits?

The challenge of identifying causal evidence for peer effects is widespread [153], commonly known as the "reflection problem" in social science [148]. Imagine two students, Allen and Mark. Allen started to hang out with Mark, the better student, and we later observe that Allen improved his grades. Did Allen do well *because* Mark had a good influence on him? Not necessarily. At least three other mechanisms could be at play:

- It is just as plausible to think that Allen chose to hang out with Mark because he was also a good student. This phenomenon is called the "selection effect": individuals generally self-select into groups of peers, making it difficult to separate out any actual peer effect.
- Allen and Mark can affect each other simultaneously, making it difficult to single out the impact Mark's outcome has on Allen's outcome.
- The observed improvement in Allen's grades can be induced by common confounding factors that we cannot observe. For example, Allen may have improved his grades simply because a new teacher or a better after-school program came along.

Given these challenges, we often rely on randomized experiments to tease out peer effects. This is the concept behind the Dartmouth dorm study, where students were assigned randomly into different rooms, thereby eliminating most alternative explanations. Alternatively, we may rely on external, unexpected shocks to a system for a so-called "natural experiment." If there is a sudden change in one parameter while other variables remain the same, we can be more certain that the response of the system is a pure response to the external shock. We will cover one such example in the next section, where researchers use the unexpected deaths of star scientists to quantify their influence on their colleagues' productivity and impact.

9.2 The Invisible College

The examples of students and cashiers suggest that peer effects are most prominent when individuals interact directly with their peers. For example, the improvement of students' GPA was only detectable at the individual room level, but absent at the dorm level [149]. In other words, an exceptional student had no influence on the performance of a neighbor down the hall. Similarly, the cashiers who became more productive were the ones who could see their star peer joining the shift. Others on the same shift who were not aware of the shift change continued at their normal pace [150]. What's particularly interesting in science, however, is that many of the observed peer effects are not limited to physical proximity. Rather, they transcend the physical space, extending into the world of ideas. To demonstrate this, let's look at what happens when superstar scientists die suddenly and unexpectedly.

In one study, researchers examined how the productivity and impact of a superstar scientist's coauthors (measured by publications, citations, and grants from the National Institutes of Health) changed when their high-performing collaborator suddenly passed away [115]. Following a superstar's death, her collaborators experienced a lasting 5 percent to 8 percent decline in their quality-adjusted publication rates. Interestingly, these effects appear to be driven primarily by the loss of an irreplaceable source of ideas rather than a social or physical proximity. Indeed, coauthors who worked on similar topics experienced a sharper decline in output than those coauthors who worked on topics further afield. And when a superstar's collaborator was heavily cited, she tended to experience a steeper decline in productivity than the superstar's less renowned collaborators. These results together provide evidence of what seventeenth-century scientist Robert Boyle called an "invisible college" of scientists, bound together by interests in specific scientific topics and ideas – a college which suffers a permanent and reverberating intellectual loss when a star is lost.

Similar effects were observed after the collapse of the Soviet Union, when the country experienced a mass emigration of Soviet scientists. Researchers studied the output of mathematicians who remained in Russia, examining what happened when their coauthors suddenly fled to other countries [154]. They found that the emigration of an *average* colleague or collaborator did not affect the productivity of those he left behind. In fact, some of his former colleagues even enjoyed a modest boost in their output, perhaps because the loss improved opportunities for those who remained. However, when researchers inspected the loss of very high-quality collaborators, measured by productivity in their fields, they uncovered major consequences. Ranking authors according to the quality of their collaborators, they found that the top 5 percent of authors suffered an 8 percent decline in publications for every 10 percent of their collaborators who had emigrated. This suggests that while we may manage the loss of an average collaborator, losing an outstanding one is, however, highly detrimental.

> **Box 9.3 The helpful scientists**
>
> Star scientists are typically defined by their individual productivity and output, such as, citations, publications, patents, and research funding [157]. But do these qualities accurately capture a scientist's value to an institution? What kind of scientist should a department hire – an investigator who churns out high-impact papers in quick succession, yet rarely finds the time to show up to department meetings? Or a researcher who works at a slower pace but who is willing to problem-solve departmental concerns, attend seminars, and offer feedback on colleagues' unpublished works?
>
> It appears that both are critical in different ways. Indeed, we have seen the significant impact that a "star" scientist can have on colleagues' productivity, and on new hires. But research shows that the "helpful" scientist also has something important, but distinct, to offer.
>
> For example, a study examining the acknowledgment sections of papers identified "helpful" scientists as those who were frequently thanked by others [158]. When these helpful scientists died unexpectedly, the quality (though not the quantity) of their collaborators' papers dropped precipitously. By contrast, the study found no such decline after the deaths of less helpful scientists. In addition, scientists who provided conceptual feedback – namely, critiques and advice – had a more significant impact on their coauthors' performance than those who provided help with material access, scientific tools, or technical work. It seems that there is often an invisible social dimension to performance. Being helpful and collegial is not merely a nice thing to do – it can genuinely affect the quality of scientific output of your colleagues.

This finding is consistent with research examining the peer effects on science professors in Germany, after the Nazi government dismissed their colleagues between 1925 and 1938 [155]. The loss of a coauthor of average quality reduced a German professor's productivity by about 13 percent in physics and 16.5 percent in chemistry. But, once again, the loss of the higher-than-average coauthors led to a much larger productivity loss. To be clear, these coauthors were not necessarily colleagues in the same university but were often located in different institutions and cities across Germany. Once again, these results speak to the idea that, at least in science, the "invisible college" is as important as the formal college in which we reside [156].

In this chapter, we have demonstrated the highly connected nature of science, showing how a scientist's productivity and creativity depends on her network of collaborators and colleagues. As much as we emphasize individual talent and celebrate independent thought in science, the fact is that scientists are highly dependent upon each other. Indeed, our productivity is strongly influenced by our departmental colleagues, regardless of whether they collaborate with us directly, or whether their research is merely related to what we do. Most importantly, we are not just influenced by our next-door neighbors. However far away our collaborators might be, their achievements can propagate through the network of ideas to produce long-lasting effects on our own careers. In other words, the idea of the lone genius was never accurate – in science we are never alone.

10 COAUTHORSHIP NETWORKS

Throughout history, only eight mathematicians have published more than 500 papers. Lucien Godeaux (1887–1975) is one of them [159]. A prolific Belgian mathematician, he's ranked fifth among the most published in history. Sitting on the very top of this list is Paul Erdős, the Hungarian mathematician we encountered in Chapter 1.

There is, however, a fundamental difference between Godeaux and Erdős. Of the 644 papers published by Godeaux, he was the sole author on 643 of them. In other words, only *once in his career* did he venture out of his lonely pursuit of mathematics and coauthor a paper with someone else. Erdős, on the other hand, is famous not only for his unparalleled productivity, but also for the more than 500 coauthors with whom he worked throughout his career. Indeed, most of Erdős' papers were the fruit of collaborations – so much so that they inspired the so-called "Erdős number," a popular pastime of mathematicians curious about their distance to this giant of mathematics.

Erdős, by definition, has an Erdős number of zero. Those who have coauthored at least one paper with Erdős are given an Erdős number of 1. Those who have coauthored with these people but not with Erdős himself have an Erdős number of 2, and so forth. It is an unparalleled honor to have an Erdős number of 1, or, in other words, to be counted among his plentiful but still rarefied list of collaborators. Short of that, it's a great distinction to be only two links away from him. In fact, having a relatively small Erdős number gives a person bragging rights not only in mathematics, but in many different disciplines – it's not unusual for scientists of all kinds to have that prized numeral listed, only partially in jest, on their CVs and websites.

The very existence of the Erdős number demonstrates how the scientific community forms a highly interconnected web, linking scientists to one another through the papers they have collaborated on. This web is often referred to as the coauthorship network. But is there a pattern underlying the way we collaborate? Which kinds of scientists are most, or least, willing to collaborate with one another? Understanding coauthorship networks, and the insights they reveal about the structure and the evolution of science, is the focus of this chapter.

10.1   A First Look at Coauthorship Network

As soon as massive digitized publication records became available around 2000, researchers started constructing large-scale coauthorship networks, capturing collaborative patterns within physics [160, 164], mathematics [161–164], biology [164], computer science [165], and neuroscience [163]. To construct a coauthorship network, we go through each publication and add a link between two scientists if they appeared on the same paper together. Figure 10.1 illustrates the local structure of one such network surrounding a randomly selected author, highlighted in the center [160]. A quick look at this network reveals several important features of collaborations: First, the network is held together by a few highly connected nodes, or hubs, who are highly collaborative individuals like Erdős. Second, the network consists of densely connected cliques of authors or communities. To highlight these communities, we can apply to this network a community-finding algorithm designed to identify cliques [160], and color the nodes based on whether they belong to a clearly identifiable clique. Those that do not belong to any recognizable community are colored black. As the figure illustrates, the vast majority of nodes take up a color, indicating that most scientists belong to at least one identifiable community.

10.2   The Number of Collaborators

With more than 500 coauthors within mathematics, Erdős is clearly an outlier. But how unusual is he for a mathematician? To find

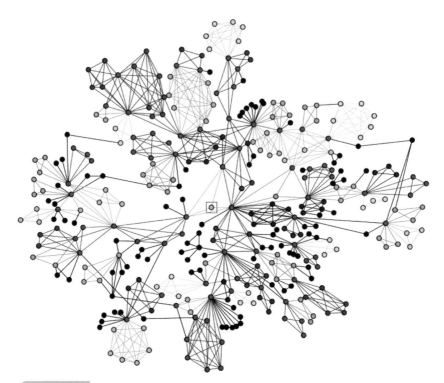

Figure 10.1 **Coauthorship network.** The figure shows the local structure of the coauthorship network between physicists in the vicinity of a randomly chosen individual (marked by a red frame). The network is constructed based on papers from Cornell University's archive server (cond-mat), the precursor of the widely used arXiv, containing at that time over 30,000 authors. Each node is a scientist, and links document collaborative relationships in the form of coauthored publications. The color-coded communities represent groups of collaborators that belong to locally densely interconnected parts within the network. Black nodes/edges mark those scientists that do not belong to any community. After Palla et al. [160].

out, next we compare the collaboration networks across biology, physics, and mathematics [164]. A key property of a node in a network is its *degree*, which represents the number of links it has to other nodes [67, 68]. In the context of the coauthorship network, the degree k_i of node i represents the number of collaborators scientist i has. The distribution of the number of collaborators that scientists have in each of the three disciplines, $P(k)$, is shown in Fig. 10.2. Each of these distributions is fat-tailed, indicating that, regardless of their discipline,

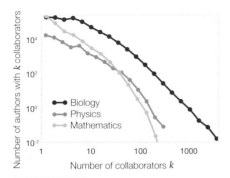

Figure 10.2 Collaboration networks are heterogeneous. The plots show the distribution of numbers of collaborators for scientists in physics, biology, and mathematics, indicating that the underlying distributions are fat tailed. After Newman et al. [164].

almost all scientists work with only a few coauthors, while a rare fraction accumulates an enormous number of collaborators. But, though all three distributions shown in Fig. 10.2 are fat-tailed, the curves follow distinct patterns. Indeed, the distribution for biology (dark) has a longer tail, reflecting the higher likelihood that a biologist will have a large number of collaborators, whereas the distribution for mathematics (green) decays the fastest among the three. Hence, highly collaborative individuals are especially rare in mathematics – and an ultra-collaborator like Erdős is rare in any discipline.

Importantly, the number of collaborators a scientist has already worked with can predict her odds of establishing new collaborations in the future. That's the idea of preferential attachment, a concept we will discuss again in Part III, stating that we are more likely to collaborate with highly collaborative individuals. Indeed, when an author publishes for the first time, odds are she'll be coauthoring her paper with senior authors, such as her advisor, someone who already has a large number of coauthors, rather than a fellow graduate student, who likely lacks collaboration links [163]. The same is true for new collaborations among scientists who are already part of the network: they are more likely to form new links to highly connected authors than less connected ones. As a consequence of preferential attachment, authors with more collaborators more quickly increase their circles of collaborators, gradually turning into hubs of the scientific collaboration network.

10.3 Small World

As word of the Erdős number spread, mathematicians around the world began to calculate their distance from the esoteric center of the math universe. Their efforts were documented by Jerry Grossman, a math professor at Oakland University in Rochester, Michigan, who maintains the Erdős Number Project [166]. If you visit the project's website, you soon realize that the Erdős number extends well beyond mathematics. Listed next to mathematicians are economists, physicists, biologists, and computer scientists who can claim a link to Erdős. Bill Gates is there, for example, thanks to his 1979 publication with Christos H. Papadimitriou, who in turn published with Xiao Tie Deng, who in turn published with Erdős' coauthor Pavol Hell. That means that Gates has an Erdős number of 4. That may sound like a small number of steps to connect a Hungarian mathematician to someone who almost never engages in scientific publishing. But, as we will see, the paths between scientists are often shorter than we might have imagined.

This kind of far-flung yet proximal connection has to do with the small-world phenomenon [167], also known as "six degrees of separation" (Box 10.1). Put in network science terms, this popular concept captures the fact that there is a short path between most pairs of nodes within a network. Indeed, if we measure the minimum number of links that separate any two scientists in a coauthorship network, the typical distance is about six links. This pattern holds for biologists, physicists, computer scientists [165], mathematicians, and neuroscientists [163]. That means that if one scientist picks another random scientist, even one she has never heard of, chances are that she will be able to connect to this person through just five or six coauthorship links. Moreover, the distance between a typical scientist and a highly connected hub of the collaboration network is even smaller. For instance, the average distance from Paul Erdős to other mathematicians is about 4.7 [166], significantly lower than the typical distance in the network as a whole.

> **Box 10.1 Six degrees of separation**
>
> The small-world phenomenon is also known as *six degrees of separation*, after John Guare's 1990 Broadway play. In the words of one of the play's characters:
>
> > Everybody on this planet is separated by only six other people. Six degrees of separation. Between us and everybody else on this planet. The president of the United States. A gondolier in Venice... It's not just big names. It's anyone. A native in a rain forest. A Tierra del Fuegan. An Eskimo. I am bound to everyone on this planet by a trail of six people. It's a profound thought.

Because research teams are embedded in the coauthorship network, the size and shape of that network can affect how teams perform. In an enormous, far-flung network, many teams will be more than six steps apart, leaving their members isolated and unable to easily exchange new ideas. On the other hand, if the world of collaborations is too small, it can form an echo chamber that is detrimental to out-of-the-box thinking. That means there may be a sweet spot in the middle which provides creative collaborators with the most advantageous environment for nurturing creative ideas.

To understand how "small-worldliness" affects a team's creativity, we can explore the collaboration patterns of the creative artists involved in the production of Broadway musicals [168]. In the Broadway network, two artists are connected if they worked together in a musical before, either as producers, directors, designers, or actors. Figure 10.3 shows three networks of teams characterized by varying degrees of the small-world property. Let's use W to denote the "small-worldliness" of each network. The map on the left of Fig. 10.3 shows a "big world" with low W where the artists are removed from one another, due to the sparseness of the links between the different teams. By contrast, the network on the right is more densely connected (high W).

Measuring each team's performance based on box-office earnings (financial success) as well as the average of the critics' reviews of their musical (artistic success), researchers found that W correlates with team performance. When a team is embedded in a low-W network, the creative artists are less likely to develop successful shows. Given the few

108 / The Science of Science

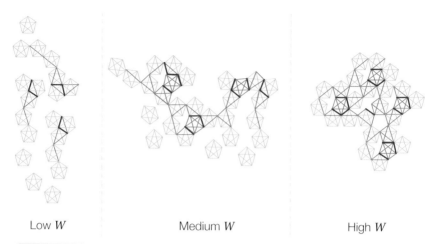

Low *W*   Medium *W*   High *W*

Figure 10.3 **Small worlds and team performance.** Diagrams of networks of Broadway artists illustrate the role of the small world effect on team performance. The parameter W quantifies the "small-worldliness" of a network. When W is low, there are few links between teams (cliques), resulting in a network of low connectivity and cohesion. As W increases, there are many between-team links, resulting in high connectivity and cohesion in the network's topology. At medium levels of W the small-world network offers an optimal amount of connectivity and cohesion. After Uzzi and Spiro [168].

links between the teams, a production team is unlikely to exchange creative ideas with many others in the network. As W increases, the network of artists becomes more connected and cohesive, facilitating the flow of creative material across clusters. This increased information flow can make the exchange of tips and conventions more likely and provide artists with feedback that empowers them to take bigger risks, boosting the performance of teams embedded in the network.

But only to a certain degree. The same study shows that too much connectivity and cohesion (high W network) can also become a liability for creativity. Indeed, cohesive cliques tend to overlook useful information that contradicts their shared understandings. Taken together, that data indicates that on Broadway, teams perform best when the network they inhabit is neither too big, nor too small.

10.4 Connected Components

The fact that Bill Gates has a small Erdős number (four) prompts a broader question: Why does Gates even have an Erdős number? After all, for Gates to have an Erdős number, there must exist a path on the coauthorship network between Gates and Erdős. If there is no such a path, then Gates would have an Erdős number of infinity. The fact that Gates has an Erdős number at all indicates that he and Erdős are on the same "connected component" of the collaboration network. In general, a *connected component* represents a group of nodes that are all connected to one another through paths of intermediate nodes. One key property of real coauthorship network is its *connectedness*; almost everyone in the community is connected to almost everyone else by some path of intermediate coauthors. For example, if we count the connected components within the three networks used in the study shown in Fig. 10.2, we will find that the largest connected component contains around 80 or 90 percent of all authors. A large connected component speaks to the idea of the invisible college, the web of social and professional contacts linking scientists across universities and continents, creating intellectual communities of shared values and knowledge base.

The fact that so many scientists are a part of the same connected component lets us draw several conclusions. First, and most obviously, it means that scientists collaborate. But if scientists always collaborate with the same set of coauthors, the network would be fragmented into small cliques isolated from each other, representing isolated schools of thought, like the low-W network seen in Fig. 10.3. Luckily, our invisible college is an expansive one. But what holds together more than 80 percent of all scientists in the same coauthorship network? Understanding the roots of this process reveals fundamental insights into how teams are assembled. To unpack these rules, let's first pay a visit to a chicken coop.

11 TEAM ASSEMBLY

William M. Muir, a professor of animal sciences at Purdue University, thinks about team productivity in a different context: chickens [169]. Specifically, Muir wanted to know whether he could group hens into different cages to maximize the number of eggs they lay. Could he, in other words, assemble highly productive teams? That question is much easier to answer with chickens than with humans, partly because Muir didn't need a hen's permission to move her around. Plus, performance in this context is easy to measure – all you need to do is to count the eggs.

Muir assembled two types of cages. First, he identified the most productive hen from each cage and put them together in the same cage, creating a select group of "super-chickens." Then Muir identified the most productive *cage* – a group of hens that had produced a lot of eggs as a team, although not necessarily as individuals – and left it intact. How much better would the offspring of the super-chickens do than the most productive existing team? To find out, he let both groups breed for six generations (a standard procedure in animal science), after which he tallied the eggs.

Six generations in, his control group of chickens from the highly productive cage were doing just fine. They were a brood of healthy, plump, and fully feathered birds, and their egg production had increased dramatically over the generations. What about the super-chickens? When Muir presented his results at a scientific conference, he showed a slide of the descendants of his carefully selected top-performers [170]. The audience gasped. After six generations, the number of chickens in the cage had dwindled from nine to three. The rest had been murdered by their coop-mates. And the once-stunning egg production of their fore-mothers was a thing of the past. The surviving descendants of the super-chickens, now sickly and pocked with battle scars, were under so much stress that they barely laid any eggs.

We assume, wrongly, that creating a successful team is as simple as recruiting the best of the best. Muir's experiment with chickens is an important reminder that there's far more to it than that. Sure, you need talented teammates, but they also must learn to work together, "to achieve something beyond the capabilities of individuals working alone" [171]. Indeed, just like recipes that produce a memorable dish, team assembly is not just about tossing the best ingredients you find in your fridge into a pot. We also need to consider how these ingredients fit together.

So how can scientists leverage singular talents without harming the greater team dynamic? Why are some teams exceptionally successful, winning in a seemingly predictable manner, while some teams flop, even when they're brimming with talent? Over the years, several reproducible patterns have emerged, which can help us understand and predict the success and failure of scientific teams.

11.1 Is There Such a Thing As "Too Much Talent"?

When it comes to humans, should we really be concerned about having too much talent on a team? We tend to think, just as Muir did, that when we group top-performers together, we'll get excellent team performance. We even have a name for this approach: the "all-star team." At the very least, an all-star team should create a critical mass, the typical thinking goes, helping us to attract even greater talents. To see if this is the case, let's look at the Duke University English department.

If there has ever been an all-star academic team, it was Duke's English department in the late 1980s and early 1990s [172, 173]. Hoping to raise Duke's profile, university administrators decided to recruit the cream of the crop. They started by hiring Frank Lentricchia, then the most famous graduate of the department, away from Rice University in 1984. Next, at Lentricchia's suggestion, the administration recruited Stanley Fish from Johns Hopkins to head the department. His chairmanship would go on to become legendary for the uncompromising and expensive recruiting efforts Fish undertook. Indeed, Fish had a knack for hiring scholars who were stars or about to become stars. In just a few years, he lured outstanding professors from everywhere: Barbara Herrnstein Smith, who was about to become president of the Modern Language Association; Guggenheim fellow Janice Radway; pioneering queer theorists Eve Kosofsky Sedgwick, Jonathan Goldberg, and Michael Moon; future African-American studies institution builder Henry Louis Gates Jr.; along with pioneering medieval literature scholar Lee Patterson and his influential wife, Annabel Patterson.

All of a sudden, Fish turned a respectable but staid English department into a *Who's Who* of the literary world. And these moves made an impact: Between 1985 and 1991, graduate applications increased four-fold. In the area of gender and literature, US News and World Report would eventually rank the graduate program first in the country. When an external review committee came in 1992, their reports were filled with nothing but admiration: "We were struck by the chorus of testimony we heard from faculty outside the department that English has become a kind of engine or life-pump for the humanities at Duke, a supplier of intellectual energy and stimulation for the university at large. It is not easy to produce changes of this sort."

One can only imagine how shocked the external review committee members were when they came back merely six years later for their routine evaluation. By the end of spring 1997, Sedgwick, Moon, and Goldberg had accepted offers elsewhere; *American Literature* editor Cathy N. Davidson had resigned her professorship to join the university administration; Lee Patterson and his wife had defected to Yale. Although Lentricchia remained at Duke, he had left the department and publicly denounced his field. Fish, the man who started Duke's empire, also made plans of his own, announcing in July that he and his wife, Americanist Jane Tompkins, were leaving for the University of Illinois at Chicago, where he would serve as dean of the College of Arts and Sciences. By then, Tompkins had practically quit teaching, working instead as a cook at a local restaurant. Alas, it seemed that professors in the all-star department had not been busy with research or teaching, but with waging ego-fueled wars against one another. The dramatic demise of the once-great department landed on the front page of the *New York Times*, sealing its tarnished legacy [172].

What unfolded in Duke's English department is often described by psychologists as the *too-much-talent effect* [174], and it applies to several other team settings as well. For example, teams composed exclusively of high-testosterone individuals experience reduced performance because group members fight for dominance [175]. Having a large proportion of high-status members can also negatively affect the performance of financial research teams [176]. Make no mistake, highly talented individuals remain critical to team success. A large body of research confirms that adding more talents to the team can help improve outcomes – but only to a certain extent, beyond which the benefits may not be guaranteed.

So, when does "all-star" become "too-much-talent"? A comparison of team performance across different sports may offer some initial answers. Using real-world sports data, researchers found that in both basketball and soccer, the teams with the greatest proportion of elite athletes performed worse than those with more moderate proportions of top players [174]. In baseball, however, extreme accumulation of top talent did not have the same negative effect. The difference may be rooted in the fact that baseball depends far less on coordination and task interdependence between team members than either basketball or soccer. This evidence suggests that the transition from "all-star" to "too-much-talent" may be particularly pronounced when teams require members to work together as a cohesive unit.

Applied to science, the implications of these findings are obvious: There are few endeavors that require more coordinated, interdependent, and harmonious teamwork than scientific collaborations. This suggests that, scientists should take care not to overstuff their teams with all-stars, lest they suffer the same fate as an overloaded soccer team or the Duke English department.

11.2 The Right Balance

Assembling a successful team is not just about finding talents, but about choosing team members who offer the right balance of traits. But how do we find the right balance? Are we better off working with people whom we are familiar with, since they "speak our language," allowing us to march forward in harmonious lockstep? Or should we pick collaborators who bring experience, expertise, and know-how quite different from our own, so that, together, we can tackle problems that none of us can solve alone? These are some of the most studied questions in the literature of team science, attracting researchers in disciplines ranging from psychology to economics to sociology [137]. Often this research hinges on the assumption that greater heterogeneity within a team leads to more diverse perspectives, which, in turn, leads to better outcomes. Yet at the same time, there are reasons to believe that diversity may not always be a good thing. For example, although diversity may potentially spur creativity, it can also promote conflict and miscommunication [177]. It therefore begs the question: Does diversity in scientific collaborations hurt or help?

Recent studies have examined many different dimensions of diversity, including nationality, ethnicity, institution, gender, academic age, and disciplinary backgrounds. These studies offer consistent evidence that diversity within a scientific team promotes the effectiveness of the team, either by enhancing productivity or resulting in works with higher impact, or both. Let us briefly review a few key findings in this domain:

Ethnic diversity: Using ethnicity as a measure of diversity, researchers found that diversity, or lack thereof, strongly influences both how much we publish and the impact of those publications [178, 179]. In general, authors with English surnames were disproportionately likely to coauthor papers with other authors with English surnames and those with Chinese names were more likely to collaborate with other Chinese scientists. The papers that result from such homophily, however, tend to land in lower impact journals and garner fewer citations. Indeed, when researchers studied the ethnic diversity of over 2.5 million research papers written by US-based authors from 1985 to 2008, they showed that papers with four or five authors from different ethnicities had, on average, 5–10 percent more citations than those written by authors all of the same ethnicity [178, 179].

International diversity: A similar effect holds for international diversity [179, 180]. Analyses of all papers published between 1996 and 2012 in eight disciplines showed that papers authored by scientists from more countries are more likely to be published by higher-impact journals and tend to receive more citations [180].

Institutional diversity: Teaming up with someone from another institution also seems to reliably improve productivity and result in highly cited work, compared to teaming up with someone down the corridor [139, 143, 179]. Examining 4.2 million papers published over three decades, researchers [143] found that across virtually all fields of science, engineering, and social science, multi-university collaborations consistently have a citation advantage over within-school collaborations.

Interestingly, across various measures of diversity, ethnic diversity of a team seems to offer the most significant lift in the impact of their papers. A group of researchers analyzed over 9 million papers authored by 6 million scientists to study the relationship between research impact and 5 classes of team diversity: ethnicity, discipline, gender, affiliation, and academic age [181]. When they plotted those diversity measures against the teams' five-year citation counts, ethnic diversity correlated with impact more strongly than did any other category (Fig. 11.1a). The researchers further showed that ethnic diversity, not some other factor, was indeed behind this impact boost. Indeed, even when they controlled for publication year, number of authors, field of study, author impact level prior to publication, and university ranking, the clear association between diversity and scientific impact remained: Ethnic diversity on a team was associated with an impact gain of 10.63 percent.

One explanation for this relationship is that scientists who collaborate with a broader variety of colleagues happen to also do better science – that is, the gain could come from having open-minded, collaborative team members, not from ethnic diversity itself. To test this explanation, researchers devised a measure called "individual diversity," which compiled all of an author's earlier coauthors and examined how ethnically diverse that list was. They then compared that to "team" diversity, the degree of ethnic diversity among coauthors of a single particular paper. The results indicate that while both team and individual diversities can be valuable, the former has a greater effect on scientific impact (Fig. 11.1b). In other words, impact is really about how diverse a team is; not how open to diversity each team member is.

Still, the mechanism behind this impact boost remains unclear. It could happen because high-quality researchers tend to attract the best people from around the world, for example, or because different

116 / The Science of Science

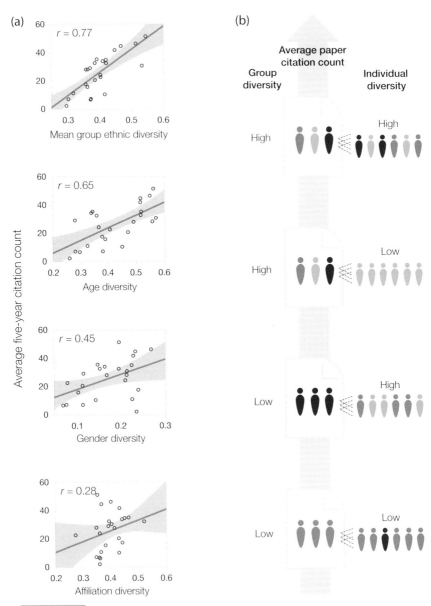

Figure 11.1 **Team diversity and scientific impact.** (a) An analysis of more than 1 million papers in 24 academic subfields (circles) shows that ethnic diversity correlates more strongly (r) with citation counts than do diversity in age, gender, or affiliation. (b) Comparing team versus individual diversity reveals that diversity within the list of authors on a paper (team diversity) has a stronger effect on citation count than diversity in a researcher's network of collaborators (individual diversity). After Powell [182].

cultures cross-fertilize ideas. Regardless of the underlying mechanisms, these findings suggest that just knowing a team's ethnic diversity can help predict their scientific impact, which in itself has implications [181]. For example, recruiters may want to encourage and promote ethnic diversity, especially if new hires complement the ethnic composition of existing members. Furthermore, while collaborators with different skill sets are often needed to perform complex tasks, these results suggest that multidisciplinarity may be just one of several critical forms of diversity. Bringing together individuals of different ethnicities – with the attendant differences in culture and social perspectives – may also pay off in terms of performance and impact.

Keep in mind, however, that the studies covered in this chapter relied on publication data, meaning that their findings are limited to teams that were *successful enough to publish* in the first place. However,

---

Box 11.1 Collective intelligence

The concept of measurable human intelligence, more commonly known as IQ, is based on one remarkable fact: People who do well on one mental task tend to do well on most others, despite large variation in the nature of the tasks [184]. While still considered controversial, the empirical fact of general cognitive ability, first demonstrated by Spearman [185], is now, arguably, the most replicated result in psychology [184]. But if we can accurately measure individual intelligence – and research shows we can – can we gauge the intelligence of a team?

The answer is provided by the concept of collective intelligence [186], referring to the general ability of a group to perform a wide variety of tasks. Interestingly, collective intelligence is not strongly correlated with the average or maximum individual intelligence of the group members. Instead, it's correlated with the average social sensitivity of group members, equality in the distribution of conversational turn-taking, and the proportion of females in the group. Teams with more socially sensitive people, fewer one-sided conversationalists, and more women have been shown to predictably exhibit superior group performance in a variety of tasks. These results suggest that a team's ability to perform well depends primarily on the composition of the team and the way team members interact, rather than the characteristics of individual team members.

it is possible that a team's diversity can increase its chances of failure in the stages before publication, potentially because of communication and coordination costs or barriers between disciplines. For example, an analysis of more than 500 projects funded by the US National Science Foundation revealed that the most diverse teams are on average the least productive – that is, diversity results in a higher risk of publishing less, or not at all [139, 183]. This highlights a major gap in our understanding of team science in general: We know very little about teams that failed, an issue that we will return to in the outlook chapter (Chapter 23).

Therefore, the insights discussed in this chapter should be interpreted under an important condition: when teams *do* form and publish, diversity is associated with a higher impact. Overall, a research team's diversity and breadth of knowledge contribute to the quality of the science that the team produces [179]. Having a diverse team means that collaborators bring different ideas and ways of thinking to the joint effort, improving the outcome of their collaborative work.

11.3 Assembling a Winning Team

How do we assemble a winning team? The most successful teams are embedded in interconnected webs of colleagues from many fields – in other words, a large, wide-ranging network from which to draw ideas. Being part of such networks can offer the kind of inspiration, support, and feedback that leads to the most winning enterprises. But not all networks are equally fertile for collaborations, and some can be less advantageous. The way a team chooses its members plays an important role in how well the team will perform. To see how, let's look at a simple model that aims to capture how teams come together [177, 187].

In any team, there are two types of members: (i) Newcomers, or rookies, who have limited experience and unseasoned skills, but who often bring a refreshing, brash approach to innovation, and (ii) incumbents, the veterans with proven track records, established reputations, and identifiable talents. If we categorize all scientists as either rookies or veterans, we can distinguish four different types of coauthorship links in a joint publication: (1) newcomer–newcomer, (2) newcomer–incumbent, (3) incumbent–incumbent, or, if both are incumbents who have worked together before, (4) repeat incumbent–incumbent.

These four types of links offer a simple way to capture the many possible configurations of diversity and expertise among team members.

For example, research on grant-funded projects demonstrates that projects whose principal investigators share prior collaboration experience (a repeat incumbent–incumbent link) are more likely to be successful than those involving members who had never written a joint paper before [183]. However, teams that predominantly consist of repeat incumbent–incumbent links may be limited in their ability to produce innovative ideas because their shared experiences tend to homogenize their pool of knowledge. In contrast, teams with a variety of links may have more diverse perspectives from which to draw but may lack the trust and shared knowledge base needed to make progress.

Varying the proportion of the four types of links within a team can help us understand how certain coauthorship patterns will impact the team's success. We can explore this relationship using two parameters [187]:

- **The incumbency parameter,** p, represents the fraction of incumbents within a team. Higher values of p indicate that a team is mostly made up of experienced veterans, whereas low values of p signal that a team consists mostly of rookies. Therefore, the incumbency parameter approximates the collective experience of a team.
- **The diversity parameter,** q, captures the degree to which veterans involve their former collaborators. An increased q indicates that incumbents are more inclined to collaborate with those with whom they have collaborated before.

By varying the two parameters, we can set up a generative model that not only captures teams with different collaborative patterns [187] (see Appendix A.1 for more details), but also helps us quantify the relationship between the team assembly patterns and innovative output. Indeed, researchers collected the publication records of teams across four different scientific disciplines (social psychology, economics, ecology, and astronomy) and extracted the incumbency and diversity parameters (p and q) by examining how teams were assembled in each case. To measure performance, they compared the values of these parameters with each journal's impact factor, a proxy for the overall quality of the team's output.

Across economics, ecology, and social psychology, there is consistent evidence that a journal's impact factor is positively correlated with the incumbency parameter p, but negatively correlated with the diversity parameter, q. This means that the teams publishing in high-impact journals often have a higher fraction of incumbents. On the other hand, the negative correlation between the journal impact factor and diversity parameter, q, implies that teams suffer when they are composed of incumbents who mainly choose to work with prior collaborators. While these kinds of team alignments may breed familiarity, they do not breed the ingenuity that new team members can offer.

Box 11.2 Successful teams

The recipe for assembling a successful team applies in a broad range of creative enterprises, from Broadway shows to video games to jazz.

Musicals: In a study of 2,258 musicals performed at least once on Broadway between 1877 to 1990 [187], researchers defined a team as the group of individuals responsible for composing the music, writing the libretto and the lyrics, designing the choreography, directing, and producing the show. And although Broadway musicals and scientific collaborations seem like vastly different creative endeavors, they share the same recipe for success. The best-received musicals benefit both from the experience of veterans and the new ideas and fresh thinking of newcomers.

Video games: A study of teams behind 12,422 video games released worldwide from 1979 to 2009 found the most effective collaborations were comprised of people who had experience working together but who brought different knowledge and skills to the task at hand [188].

Jazz: Data on the entire history of recorded jazz from 1896 to 2010 (175,000 recording sessions) revealed that combining musicians who have worked together with new players is critically important to the success of a jazz album, as measured by the number of times an album is released [189]. That was certainly the case with Miles Davis' *Kind of Blue*, the most re-issued jazz album in history. By the time the album was recorded, Paul Chambers, the bass player, was a true veteran who'd played a total of 58 sessions. Out of these sessions, 22 were with Miles Davis, the trumpeter and band leader, and 8 were with Wynton Kelly, the pianist. However, Davis and Kelly had never played together prior to *Kind of Blue*. Therefore, while Davis and Kelly were both deeply familiar to Chambers, the fact that they did not know each other added a fresh dynamic to the powerful linkages the incumbent team members already shared, creating a situation ripe for innovation.

Taken together, these results show that there's a strong correlation between team composition and the quality of the work they produce. In particular, these findings teach us two important lessons about how to assemble a successful team [177]. First and foremost, experience matters: While brash newcomers may brim with energy and are fearless in taking risks, teams composed entirely of rookies are more likely to flop. Second, although the experience of veteran scientists is important, successful teams include some portion of new people who can stimulate new ideas and approaches. Too many incumbents, especially those who have worked together repeatedly in the past, limit diverse thinking, which can lead to low-impact research.

11.4 The Dynamic Duos

If team success depends on the balance of new collaborators and existing ones, how do scientists typically engage with their coauthors? Do we tend to gravitate towards a few very close collaborators, or do scientific collaborations feature a high turnover rate? An empirical analysis of 473 researchers from cell biology and physics and their 166,000 collaborators reveals three important patterns in the way we form and maintain collaborations [190].

First, scientific collaborations are characterized by a high turnover rate, dominated by weak ties that do not last. Indeed, out of the more than 166,000 collaboration ties, 60–80 percent lasted for only a single year. And even when a relationship lasted more than two years, it didn't have much long-term staying power: Roughly two-thirds of collaborators went their separate ways within five years.

Second, whereas weak ties dominate most collaborations, "super ties" – extremely close working relationships – are more common than expected. Nine percent of biologists and 20 percent of physicists have a super-tie collaborator with whom they coauthored more than half of their papers. In particular, 1 percent of collaborations lasted more than 20 years. These capture lifelong collaborative partners – they are the Batmen and Robins of science. When we quantify the frequency of super ties [190], we find that they occur every 1 in 25 collaborators on average.

Third, and most important, super ties yield substantial productivity and citation premiums [190, 191]. If we compare the papers a scientist published with her super tie to those she published with other collaborators, her *productivity with the super tie was roughly eight times higher*. Similarly, the additional citation impact from each super tie is 14 times larger than the net citation impact from all other collaborators. For both biology and physics, publications with super ties receive roughly 17 percent more citations than their counterparts. In other words, the work you do with your super tie, if you have one, is what tends to define your career.

The significant positive impact of super ties reflects the fact that "life partners" in science exist for a reason. For example, researchers analyzed collaboration records among 3,052 Stanford University faculty between 1993 and 2007, extracting data from student dissertation records, grant proposal submissions, and joint publications [191]. They identified when new, untenured faculty arrived on campus, tracing when these newcomers began to form their first ties and exploring why some of these relationships persisted over time, becoming repeat collaborations. They found that repeated collaborations are of a fundamentally different nature from first-time collaborations – whereas new collaborations are mostly driven by opportunity and preference, repeated collaborations are a function of obligation and complementary experience. Indeed, when someone seeks a collaborator for the first time, they tend to select someone who has a similar skill-set, as they are less familiar with the other person's workstyle. Yet, as the two work together on more projects, they become better acquainted with each other's unique talents. If they decide to continue collaborating, it suggests that each individual has realized that their partner has something novel to contribute, which may lead to a better final outcome. Thus, productive relationships are most often sustained when the individuals have non-identical knowledge, so they can be complementary to each other.

The value of a super tie is exemplified by the case of Michael S. Brown and Joseph L. Goldstein. Professors at the University of Texas Southwestern Medical School, Brown and Goldstein have both enjoyed stellar careers and achieved remarkable productivity, each publishing over 500 papers and earning numerous awards, including the Nobel Prize and National Medal of Science. Yet a quick look at their publication records reveals a truly special relationship: More than 95 percent of their papers are jointly authored. Searching through the Web of Science database, we find that Brown and Goldstein coauthored a stunning 509 papers by 2018. Over the course of their long careers, Goldstein has published only 22 papers without Brown, and Brown has published only 4 papers without Goldstein.

The impact of their joint research goes way beyond their academic accolades. The duo has made real impact in improving human health. In a joint paper published in 1974, Brown and Goldstein discovered LDL (low-density lipoprotein) receptors, which earned them a Nobel Prize in Physiology or Medicine in 1985 [192]. Low-density lipoprotein, which is often referred to as the "bad cholesterol," is highly associated with cardiovascular diseases like coronary artery disease, the number one killer of people worldwide [193]. Drugs that help lower cholesterol levels, called statins, are more widely prescribed than any other type of medication. Atorvastatin, the most popular variant of statins (marketed under the brand name Lipitor) is the world's best-selling drug of all time, with more than US$125 billion in sales over its 14-year life-span. It's not a stretch to say, then, that millions of lives have been saved thanks to Brown and Goldstein's super tie.

12 SMALL AND LARGE TEAMS

In 2015, *Physical Review Letters* (PRL), one of the most prestigious publication venues for physicists, published a highly unusual paper [194]. *Physical Review Letters*' stated goal is to offer a venue for rapid communication of important results – hence each paper has traditionally been a short "letter" limited to four pages in length. The 2015 paper, on the other hand, was a record-breaking 33 pages long. Yet only 9 pages featured the research itself; the remaining 24 were used to list the authors and their institutions. This high-energy physics paper was the first joint paper produced by the teams that operate ATLAS and CMS, two massive projects conducted using the Large Hadron Collider (LHC) at CERN, Europe's particle-physics lab near Geneva, Switzerland. Together, an astonishing 5,154 authors contributed to the discovery, representing the largest number of contributors to a single research article in history [195].

One of the most profound shifts in science and technology today is the shift toward larger and larger teams across all fields of science. In 1955, the average team size in science and engineering was around 1.9, indicating that two-person partnerships were the norm for collaborative work. This number nearly doubled over the next 45 years, corresponding to the 17 percent growth in team size per decade [104, 136] (Fig. 12.1). And this growth continues even today: While in 2000 a typical paper had 3.5 authors, by 2013 that number had grown to 5.24. Similar (though less dramatic) growth has occurred in the social sciences: The average social science paper today is written by a pair of authors rather than solo authors, with the number of collaborators

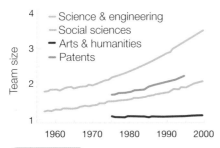

Figure 12.1 **The growing size of teams.** Each line represents the arithmetic average of team size taken over all subfields in each year. After Wuchty et al. [136].

continuing to grow every year. Even in the arts and humanities, where over 90 percent of papers are still single authored, there is also a significant, positive trend toward team-authored papers ($P < 0.001$). Furthermore, increases in team size are not unique to academia, but apply to other creative work as well. For example, in the US, the average number of inventors on patent applications has risen from 1.7 to 2.3, with the number continuing to grow every year.

This trend is about more than just simple growth. As we will see in this chapter, it reflects a tectonic shift in the ways that scientists work and teams organize themselves. And that change has tremendous consequences for the future of science, since large and small teams produce fundamentally different, though equally valuable, types of research.

12.1   Not Simply a Change in Size

The shift toward larger teams in science reflects a fundamental change in team composition [196]. Figure 12.2 compares the distributions of team size in astronomy in two periods: 1961–1965 and 2006–2010. The average team size in astronomy grew from 1.5 to 6.7 between these two periods (see the arrows in Fig. 12.2). Yet the 2006–2010 team size distribution is not simply a scaled-up version of the 1961–1965 distribution. Rather, the two distributions have fundamentally different shapes. In the 1961–1965 period, the number of papers decays rapidly as the team size increases. Not only is the average team size small, but we find not a single paper with more than eight authors. Indeed, the team sizes in the 1960s are best approximated with an exponential distribution (blue curve in Fig. 12.2), meaning that most

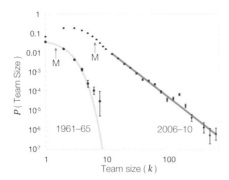

Figure 12.2 Fundamental changes in team structure. The team size distribution between 1961 and 1965 is well described by a Poisson distribution (blue curve). In contrast, the 2006–2010 distribution features an extensive power-law tail (red line). The arrows mark the mean values of each distribution. Error bars correspond to one standard deviation. After Milojević [196].

teams have sizes within close vicinity of the average, and that there are no outliers.

Yet, after 45 years, the distribution has changed dramatically, and now features a prominent tail, consisting of mega-teams with several hundred authors. Unlike in the 1960s, the tail of this more recent distribution is well approximated by a power law function (red line in Fig. 12.2). This shift from an exponential to a power law distribution implies that the increase in a team's size is more complicated than it may seem. Indeed, these two mathematical functions indicate that two fundamentally different modes characterize the process of team formation.

The first and more basic mode leads to the creation of relatively small "core" teams. In this mode, new members join the team with no regard for who is currently on the team. Core teams thus form by a purely random process and thus produce a Poisson distribution of team size, making large teams exceedingly rare. As such, the distribution is best depicted with an exponential function. This mode was likely the status quo in astronomy back to the 1960s.

The second mode produces "extended" teams. These start as core teams but accumulate new members in proportion to the productivity of their existing members – a "rich-get-richer" process. As such, a team's current size influences its ability to attract new members. Given time, this mode gives rise to a power-law distribution of team size, resulting in the large teams with 10–1,000 members that we observe in many fields today.

A model of team formation based on core and extended teams can accurately reproduce both the empirically observed team size distributions and their evolution over time in many fields [196]. Most important, by fitting the model to a particular discipline across several different eras, it allows us to assess how the two modes of team formation have affected knowledge production. Interestingly, the results indicate that the fraction of articles produced by core and extended teams has remained roughly constant over time. This means the increase of team size is not because larger teams have replaced the smaller core teams. Rather, we observe a growth in both types of teams – a gradual expansion of core teams and a faster expansion of extended teams.

These changes in team formation may also have consequences for a team's longevity, as team size may be a crucial determinant of survival [160]. Research on the lifetime of collaborative teams suggests that large teams tend to persist for longer if they dynamically alter their membership, as changing their composition over time may increase their adaptability. On the other hand, small teams tend to be more stable over time if their membership remains more stable; a small team with high turnover among its members tends to die out quickly.

So far, we've demonstrated that in science, teams are growing in size, hence innovation increasingly happens in larger team settings. Is that a good thing for scientists – and for the creation of knowledge in general?

12.2 Team Size: Is Bigger Always Better?

The LIGO experiment, which offered the first evidence of gravitational waves, was called the "discovery of the twenty-first century," and recognized by the Nobel prize within two years of discovery. This experiment – the collective work of more than a thousand researchers – is a testament to the power of large teams in tackling the twenty-first century's toughest challenges. Indeed, one could argue that the shift toward large teams fulfills an essential function: Since the problems of modern society are increasingly complex, solving them requires large, interdisciplinary teams [7–10] that combine their members' talents to form an enormous and varied pool of expertise. Furthermore, the LIGO experiment exemplifies the kind of achievement that simply is not feasible for smaller groups to pull off. The project required unprecedented technology, and thus demanded unprecedented material and personnel resources. It is no surprise that the paper reporting the discovery listed more than 1,000 researchers.

The universal shift toward larger and larger teams across science and technology suggests that large teams, which bring more brain-power and diverse perspectives, will be the engines for tomorrow's largest advances. Research has consistently shown that as teams become larger, their products – be they scientific papers or patented inventions – are associated with higher citation counts [136, 197]. These trends seem to offer a simple prescription for the future of science: bigger is always better.

Yet there are reasons to believe that large teams are not optimal for all tasks. For example, large teams are more likely to have coordination and communication issues – getting everyone onboard to try out an unconventional hypothesis or method, or convincing hundreds of free-thinking individuals to change direction at once, is often challenging. Psychology research shows that individuals in large groups think and act differently. They generate fewer ideas [198, 199], recall less learned information [200], reject external perspectives more often [201], and tend to neutralize one another's viewpoints [202]. Large teams can also be risk-averse, since they have to produce a continuous stream of success to "pay the bills" [203].

All of which raises the question of whether relying on large teams is truly a one-size-fits-all strategy for producing groundbreaking science. Indeed, new evidence suggests that team size fundamentally dictates the nature of work a team is capable of producing, and that smaller team size confers certain critical benefits that large teams don't enjoy.

12.3 Large Teams Develop Science; Small Teams Disrupt It

To understand how team size may affect the nature of the science and technology they produce, let's consider two examples. In Fig. 12.3, we pick two well-known papers with similar impacts, but which contribute to science in very different ways. The Bak, Tang, and Wiesenfeld (BTW) article on self-organized criticality [204] received a similar number of citations to the Davis et al. article on Bose–Einstein

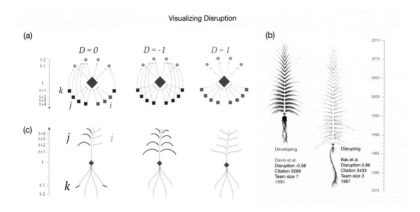

Figure 12.3 Quantifying disruption. (a) Citation network depicting a paper (blue diamond), its references (gray circles), and subsequent works (rectangles). Subsequent works may cite: (1) only the focal work (i, green), (2) only its references (k, black), or (3) both the focal work and its references (j, brown). Disruption of the focal paper suggests that a paper may either balance disruption and development (*Disruption* = 0), primarily broadcast the importance of prior work (*Disruption* = −1), or completely overshadow prior work by receiving all subsequent attention itself (*Disruption* = 1). (b) Citation tree visualization that illustrates how a focal paper draws on past work and passes ideas onto future work. "Roots" are references cited by the focal work, with depth scaled to their publication date; "branches" on the tree are citing articles, with height scaled to publication date and length scaled to their number of future citations. Branches curve downward (brown) if citing articles also cite the focal paper's references, and upward (green) if they ignore them. (c) Two articles of similar impact are represented as citation trees, "Bose–Einstein condensation in a gas of sodium atoms" by Davis et al., and "Self-organized criticality: An explanation of the 1/f noise" by Bak et al. After Wu et al. [206].

condensation [205]. Yet the Bak et al. paper was groundbreaking in a way that the Davis et al. paper was not. Indeed, most research following upon Bak et al. cited the BTW paper only, without mentioning its references (green links in Fig. 12.3a). By contrast, Davis et al., for which Wolfgang Ketterle was awarded the 2001 Nobel Prize in Physics, is almost always cocited with the papers the Davis et al. paper itself cites (brown links in Fig. 12.3a).

The difference between the two papers is not reflected in citation counts, but in whether they develop or disrupt existing ideas – i.e., whether they solve an established scientific problem or raise novel questions. "Developmental" projects, those that build upon earlier research, seek to advance understanding of an existing problem, hence they tend to be cited along with earlier work, like the Davis et al. paper. "Disruptive" projects, on the other hand, tend to establish a brand new frontier of inquiry, hence they are more likely to be cited without the predecessors, since they represent a departure from previous knowledge.

So, do large or small teams create the more disruptive work?

To quantify the degree to which a work amplifies or eclipses the prior art it draws upon, we use a disruption-index D, which varies between -1 (develops) and 1 (disrupts) [207]. For example, the BTW model paper has a disruption index of 0.86, indicating that most of the papers that cite it, do so by citing the BTW paper alone, ignoring the work that the BTW paper built upon. By contrast, the Davis et al. paper has a $D = -0.58$, indicating that it is frequently cocited with its predecessors. This reflects the fact that the BTW model launched entirely new streams of research, whereas Davis et al.'s experimental realization of Bose–Einstein condensation, while a Nobel-worthy effort, primarily elaborated on possibilities that had already been posed elsewhere.

Our analysis shows that over the past 60 years, larger teams have produced research articles, patents, and software products that garner higher impact than smaller teams. Yet, interestingly, their disruptiveness dramatically and monotonically declines with each additional team member (Fig. 12.4). Specifically, as teams grow from 1 to 50 members, the disruptive nature of their papers, patents, and products drops by 70, 30, and 50 percentiles, respectively. These results indicate that large teams are better at further developing existing science and technology, while small teams disrupt science by suggesting new problems and opening up novel opportunities.

But is the observed difference in the work produced by large and small teams really due to team size? Or can it be attributed to differences in other confounding factors? For example, perhaps small teams generate more theoretical innovations, which tend to be more disruptive, and large teams generate more empirical analyses, which are more likely to be developmental. Or, maybe there are differences in the topics that small and large teams tend to tackle. Another possibility: Perhaps certain types of people are more likely to work for smaller or larger teams, thus changing the outcomes associated with each.

Luckily, we can control for each of these plausible factors, finding that the effect of team size appears to arise as a result of team

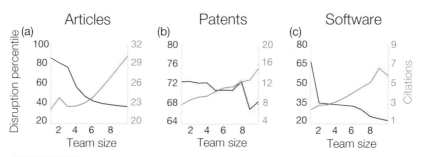

Figure 12.4 **Small teams disrupt, big teams develop.** To examine the relationship between team size and disruption, we collected data from three domains: (1) the Web of Science database, containing articles and the papers they cite; (2) patents granted by the USPTO and their citations; (3) software projects on GitHub, a web platform that allows users to collaborate on software and "cite" other repositories by building on their code. For research articles (a), patents (b), and software (c), median citations (red curves indexed by the right y-axis) increase with team size, whereas average disruption percentile (green curves indexed by the left y-axis) decreases with it. Teams of between 1 and 10 authors account for 98% of articles, 99% of patents, and 99% of code repositories. Bootstrapped 95% confidence intervals are shown as gray zones. After Wu et al. [206].

dynamics, rather than because of qualitative differences between individuals in different-sized teams. Differences in topic and research design may only account for a small part of the relationship between team size and disruption. Most of the effect (66 percent of the variation) indeed appears to be a function of team size.

Do large and small teams turn to different sources when conducting research? To answer this question, we measured how deeply small and large teams build on past literature by calculating the average age of references cited. We find that solo and small teams were much more likely to build on older, less popular ideas. This is likely not a function of knowledge: Since larger teams have more members, their expertise spans a broad range of subjects; as such, their members were probably just as aware of older, less known work as the scientists who work within small teams. However, large-team scientists tend to source their ideas from more recent, and higher-impact, work. Consequently, large teams receive citations quickly, as their work is immediately relevant to contemporaries. By contrast, smaller teams experience a much longer citation delay, but their work tends to persist further into the future, achieving a more enduring legacy.

12.4 Science Needs Both Large and Small Teams

Together, the results in the last section demonstrate that small teams disrupt science and technology by exploring and amplifying promising ideas from older and less popular work, whereas large teams build on more recent results by solving acknowledged problems and refining existing designs. Therefore *both* small and large teams are crucial to a healthy scientific ecosystem.

Large teams remain as an important problem-solving engine for driving scientific and technological advances, especially well-equipped to conduct large-scale work in personnel- and resource-intensive areas. The LIGO experiment, for example, was a feat that no small team could have achieved and has opened up a new spectrum of astronomical observation. But, while the successful detection of gravitational waves required a large and diverse team, it is also important to note that the theoretical framework that LIGO was designed to test was proposed by a single individual [208]. One kind of team was required to propose the concept of gravitational waves, and a very different kind of team was required to detect them. Both endeavors moved science forward, but in very different ways.

What do these results mean for us scientists? When putting together a team to tackle a particular challenge, we may want to first consider the goals of the project. Small teams are more agile, and better positioned to test a novel or disruptive idea. This means one may be better off starting with a smaller team in the proof-of-concept phase and engage the larger team later to fulfill the idea's promise. For scientists, it may be easy to believe that adding another member or three to a team will always be the right choice, or at the very least can't hurt. But as the findings in this chapter show, when trying to develop innovative ideas, more people isn't always better, as larger teams shift the focus and outcome away from disruption.

It is also important to think carefully about funding decisions in light of these findings. Naturally, we can count on large teams to reliably produce important work. Yet we can't assume that the bigger the team, the better its contribution will be. There is evidence that funding agencies may prefer larger teams over smaller ones, even when both are equally qualified [209]. This tendency can contribute to a self-fulfilling prophecy, funneling support disproportionately to large teams, and promoting a scientific world wherein those that master highly collaborative work have a crucial advantage.

However, as this chapter shows, there is a key role in science for bold solo investigators and small teams, who tend to generate new, disruptive ideas, relying on deeper and wider reach into the knowledge base. Hence, as large teams flourish in an environment historically populated with small teams and solo investigators, it is crucial to realize that this shift also implies that there are now fewer and fewer small teams. If this trend continues, the scientific community may one day find itself without the scrappy, outside-the-box thinkers who supply large teams with grandiose problems to solve.

13 SCIENTIFIC CREDIT

At the Nobel Prize banquet in December 2008, Douglas Prasher, dressed in a white tie and tails, sat with his wife, Gina, beneath glittering chandeliers suspended from the Blue Hall's seven-story ceiling. Prasher was in Stockholm to celebrate the impact his work has had on the world: Green fluorescent protein (GFP), the luminous protein that Prasher had cloned for the first time in 1992, had become "a guiding star for biochemistry," according to the Nobel Foundation, allowing scientists to glimpse the inner workings of cells and organs in unprecedented detail.

However, Prasher was attending the banquet as a *guest*, not a prizewinner. Among the honorees that night, the celebrated winners of the 2008 Nobel Prize in Chemistry, were Martin Chalfie and Roger Tsien, two researchers to whom Prasher had mailed his cloned GFP gene in a moment of career-crushing despondency just when he decided to leave science. On the morning of October 8, 2008, when Prasher heard on his kitchen radio the news that GFP-related research had won the Nobel Prize, he was just getting ready to go to work – not in a lab or university, but as a courtesy van driver at a Toyota dealership in Huntsville, Alabama. Prasher's trip to Sweden, paid for by the newly minted prizewinners, was the first vacation he had taken in years. The tuxedo he wore was rented for the night, and it paired well with the dressy shoes that a Huntsville store had let him borrow. He watched the ceremony from the audience because his decision to leave science decades ago also dropped him off the Nobel committee's map.

In this and the next chapters, we are going to focus on a question of growing importance today: How is scientific credit allocated? Indeed, for the solo-authored work that dominated previous eras, there was no ambiguity as to where the credit should go. But since modern discoveries are increasingly made by teams, involving anywhere between two and a *thousand* authors, who should we attribute the credit to? Or perhaps the more pragmatic question isn't who *should* get the credit, but who *will* get the credit? In this chapter, we will examine the first question; in the next chapter, we will address the second.

For many, discussions about credit are unnerving, even taboo. Indeed, we do science for the sake of science, not to revel in the glory of recognition. But, whether we like it or not, credit is allocated unevenly, so it is important to understand how this process works, if not for ourselves, then for our students and collaborators. Especially now that we have learned so much about team science, we wouldn't want the issue of credit to stand in the way of very valuable collaborations. While none of us want to claim credit that doesn't belong to us, we also don't want to find ourselves in Douglas Prasher's borrowed shoes at the Nobel Prize banquet, clapping along in stunned disbelief while someone else accepts the trophy for his work.

13.1 Who Did What in a Paper?

Compared with other professions, science is notorious for its emphasis on celebrating individuals rather than teams, particularly when it comes to rewards and recognition [105]. Indeed, iconic achievements are often known by the discovering scientist's name: Euclidean geometry, Newton's laws of motion, Mendelian inheritance, and the Heisenberg uncertainty principle, to name a few. Similarly, the science prizes that bestow significant financial reward and notoriety – from the Nobel Prizes, to the Fields Medal, to the A. M. Turing Award – tend to exclusively value individual contributions.

While the mode of production in science has been shifting toward teams, important decisions in science are still based almost entirely on individual achievement. Indeed, appointment, promotion, and tenure processes in academia center on individual evaluations, despite the fact that most careers are now built through teamwork. We scientists therefore often need to distinguish our contributions from our collaborators', whether applying for a grant, being considered for

136 / The Science of Science

(a) VOLUME 76, NUMBER 11   PHYSICAL REVIEW LETTERS   11 MARCH 1996

Generation of Nonclassical Motional States of a Trapped Atom

D. M. Meekhof, C. Monroe, B. E. King, W. M. Itano, and D. J. Wineland
Time and Frequency Division, National Institute of Standards and Technology, Boulder, Colorado 80303-3328

(b) VOLUME 55, NUMBER 1   PHYSICAL REVIEW LETTERS   1 JULY 1985

Three-Dimensional Viscous Confinement and Cooling of Atoms by Resonance Radiation Pressure

Steven Chu, L. Hollberg, J. E. Bjorkholm, Alex Cable, and A. Ashkin
AT&T Bell Laboratories, Holmdel, New Jersey 07733

(c) VOLUME 61, NUMBER 21   PHYSICAL REVIEW LETTERS   21 NOVEMBER 1988

Giant Magnetoresistance of (001) Fe/(001) Cr Magnetic Superlattices

M. N. Baibich,[a] J. M. Broto, A. Fert, F. Nguyen Van Dau, and F. Petroff
Laboratoire de Physique des Solides, Université Paris-Sud, F-91405 Orsay, France

P. Eitenne, G. Creuzet, A. Friederich, and J. Chazelas
Laboratoire Central de Recherches, Thomson CSF, B.P. 10, F-91401 Orsay, France

Figure 13.1 Who gets the Nobel? (a) The last author, David J. Wineland was awarded the 2012 Nobel Prize in Physics for his contribution to quantum computing. (b) Steven Chu, the first author, won the 1997 Nobel in Physics for the paper focusing on the cooling and trapping of atoms with laser light. (c) In 2007, Albert Fert, the middle author of the paper, received the Nobel Prize in Physics for the discovery of the giant magnetoresistance effect (GMR). All three examples are prizewinning papers published in the same journal, *Physical Review Letters*, demonstrating the ambiguity of allocating credit by simply reading the byline of a paper.

an award, seeking an academic appointment, or requesting a promotion. So how do tenure committees and award-granting institutions attempt to discern which individuals deserve credit for collaborative research?

The difficulty with allocating credit for collaborative work is rooted in a deep information asymmetry: We cannot easily discern the significance of individual contributions by reading the byline of a paper. For example, Fig. 13.1 shows three Nobel prizewinning papers with multiple authors, each published in the same journal, *Physical Review Letters (PRL)*. Only one author from each paper was awarded the Nobel Prize. So, who were the lucky prizewinners? As the figure shows, there is no simple answer: the Nobelist is sometimes the first author, sometimes the last, or can be somewhere in between.

To be clear, this issue isn't just about recognition, or giving credit where credit is due; it's also about responsibility – because with greater contribution comes greater responsibility. Who is responsible for which part of the project? Lacking consistent guidelines for classifying individual contributions to a paper, we rely on our own assumptions, traditions, and heuristics to infer responsibility and credit. The most commonly used method is by inspecting the order of authors on a paper [51, 210]. In general, there are two norms for sorting authors, alphabetical and non-alphabetical. Let's start with the more prevalent one: non-alphabetical ordering.

13.2 The First or the Last

In many areas of science, especially the natural sciences, authors order themselves on a byline based on their contributions within the team. In biology, the individual performing the lion's share of the work is the lead author, followed sequentially by those making progressively lesser contributions. Therefore, we expect that the second author had less substantial contributions than the first, but more than the third author, and so on. The critical exception is the last author, who often gets as much credit, if not more, than the first author. Listing the principal investigator last has become an accepted standard in natural sciences and engineering. The last author, also called the corresponding author, is assumed to be the intellectual, financial, and organizational driving force behind the research. Hence evaluation committees and funding bodies often take last authorship as a sign of intellectual leadership and use this as a criterion in hiring, granting, and promotion.

But does the order of authorship indeed reflect what it purports to represent? We now have quantitative answers to this question, thanks to specific author contribution sections included in publications across multiple disciplines [211–213]. *Nature*, for example, started to include explicit lists of author contributions in 1999 [214], a practice now adopted by many other journals. While the exact classification scheme differs from journal to journal, authors are generally asked to self-report their contributions in six categories: (1) analyzed data, (2) collected data, (3) conceived experiments, (4) performed experiments, (5) wrote the paper, and (6) revised the paper. Analyzing these self-reports allows us to relate the duties performed on any given project to author order. Indeed, by analyzing about 80,000 articles published between 2006 and 2014 in the multidisciplinary journal *PLOS ONE*, researchers confirmed that first and last authors

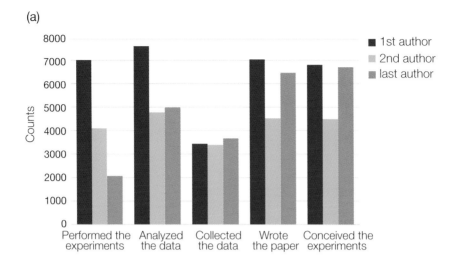

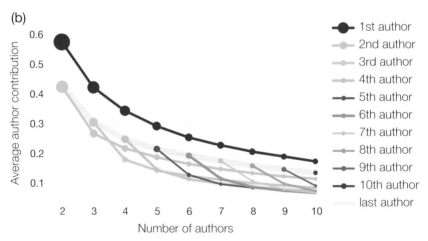

Figure 13.2 **Author contributions in a paper.** (a) Authors of different ranks make different contributions to the paper. (b) Average number of distinct contributions by an author as a function of its ranking, confirming that typically the first and last authors contributed most to the paper. After Corrêa et al. [211].

are indeed the two most important on a paper: The first author is typically involved in almost all aspects of the reported research, from running the experiment to writing the paper, and the last author assumes a supervisory role for the project, mainly designing and conceiving the research and writing the paper (Fig. 13.2a). Regardless of how many authors are on the paper, the first author always makes the largest number of contributions, and the last author usually contributes the second most (Fig. 13.2b). Interestingly, however, only the first and the last authors stand out – followed by the second author but with almost no detectable difference in contributions for all other authors on the paper. The contributions by the third and fourth author, for instance, are typically indistinguishable (Fig. 13.2b).

These results suggest that author positions in the byline can be an effective way to infer each team member's true contributions. But, there are caveats. First, the way that journals collect this information is not exact. Since there is no clear definition of each contribution, just general criteria, authors may have their own interpretations about what "analyzed data" or "conceived experiments" means. And it seems that, at times, coauthors do not necessarily agree amongst themselves about who contributed what. For example, in a study of all manuscripts submitted to the *Croatian Medical Journal*, researchers compared which contributions the corresponding author credited to their coauthors versus which contributions those authors claimed for themselves. They found that more than two-thirds (69.4%) of 919 corresponding authors disagreed with their coauthors regarding contributions [215]; specifically, coauthors listed more contribution categories on their own forms than the corresponding authors chose for them.

Second, the practice of listing authors based on their contributions is not always exercised [51]. Sociology and psychology, for example, do not follow the tradition of putting the most senior author last. In these disciplines, being the last author really *is* the least desirable position. Moreover, it is not uncommon for a mentor to be in the first-author slot for the crucial guidance, intellectual contributions, and financial leadership of the project, even if the trainee may have done all the heavy lifting.

The uneven acceptance of senior authorship often leads to ambiguities when evaluating a scientist, especially when it comes to cross-disciplinary hires. A psychology professor may ask why a prospective applicant from another discipline doesn't have any recent first-author publications, since in psychology, the absence of first-authorship is something to be frowned upon, even for an established scientist. But since listing senior authors last is the norm in other fields, like biology and physics, a biologist or physicist may see too many first-author papers as a red flag, calling into question the candidate's collaboration style and mentorship abilities.

The different preferences on author ordering also sometimes lead to conflicts. For example, a 2017 paper published in *Genes, Brain, and Behavior* [216] was retracted a few months after its publication, but not for academic misconduct or error, but due to a dispute over the order of the authors, prompting the journal to release an unusual statement:

> The retraction has been agreed as all authors cannot agree on a revised author order, and at least one author continues to dispute the original order.

Box 13.1 Joint first authorship

The percentage of publications in which two or more coauthors claim joint first authorship has dramatically increased in the past 30 years. Such "co-first authorship," as it's called, is particularly prominent in biomedical and clinical journals. Many journals went from having no such papers in 1990 to having co-first authorship in more than 30 percent of all research publications in 2012 [217].The increase has been most dramatic in high-impact journals: In 2012, 36.9 percent of *Cell* papers, 32.9 percent of *Nature* papers, and 24.9 percent of *Science* papers had co-first authors.

There are many reasons for the increasing prevalence of joint first authorships. First, publishing in a competitive journal requires extensive effort by more than one lead investigator. Hence this trend is likely to continue, especially as research grows increasingly complex and as teams are increasingly needed to tackle important scientific and clinical challenges. Also, in interdisciplinary work, a project often requires the joint, focused contribution of two or more lead authors with complementary expertise; no one individual has all of the knowledge needed to usher a project through independently.

Yet, co-first authorships also introduce new credit allocation problems. Indeed, although co-first authorship is clearly indicated by asterisks or superscripts in the original paper, when a citation is encountered on a reference list, there is no way of knowing whether it was a co-first author situation or not. Worse still, for long author lists, citations adopt the "et al." style, which lists the first author's last name only. At the time of this writing, we are only just beginning to develop new conventions around joint authorships: Journals like *Gastroenterology* [218, 219] and *Molecular Biology of the Cell* [220], for instance, now require authors to use bold lettering in citations referring to joint first authorships, so that every deserving author receives equal recognition every time their work is cited.

13.3 From A to Z

While biology or medicine place special emphasis on first and last authors, disciplines like mathematics and experimental particle physics are known for their stubbornly alphabetical approach to author lists. Yet when researchers estimated alphabetical authorship across all fields of science, they found that the overall use of the practice is on the decline [221]. In 2011, less than 4 percent of all papers chose to list their authors alphabetically. Admittedly, 4 percent is a small number, but that does not mean that alphabetical authorship is likely to disappear entirely. The decline simply reflects the fact that non-alphabetical disciplines, like biology and medicine, are expanding more rapidly than fields like mathematics. Indeed, the alphabetic convention remains highly concentrated in a handful of scientific subjects, within which its use has remained rather stable. Alphabetic listing of authorship is most common in mathematics, accounting for 73.3 percent of papers (Table 13.1). Finance comes in second with 68.3 percent, followed by economics and particle physics.

Why don't these fields embrace a more informative non-alphabetical convention? A statement from the American Mathematical Society (AMS) offers a hint [166]:

> In most areas of mathematics, joint research is a sharing of ideas and skills that cannot be attributed to the individuals separately. Researchers' roles are seldom differentiated (as they are in laboratory sciences, for example). Determining which person contributed which ideas is often meaningless because the ideas grow from complex discussions among all partners.

Additionally, in fields that obey alphabetical ordering, the definition of authorship is substantially different from those that don't. In non-alphabetical disciplines, since the placement of names within the author list clearly suggests the role they played in the project, individuals who made a marginal contribution to the work can be added to the author list in a way that does not devalue the core authors' hard work. But since everyone on an alphabetical list is considered equal, adding an undeserving person is far more costly, and hence warrants more delicate considerations. For example, in alphabetical fields like economics, merely contributing data to a project doesn't warrant authorship, while in biology or medicine, which are non-alphabetical disciplines, it frequently does. Consequently, the acknowledgments section on a mathematics paper often contain acknowledgements for significant "author-worthy"

Table 13.1 **The prevalence of alphabetical ordering of authors.** The table shows the fraction of "alphabetical" papers in a given subject, listing a selection of subjects that have a low, mid, or high fraction of intentionally alphabetical papers. Intentionally alphabetical ordering is estimated by comparing the observed number of alphabetical papers with the expected number of coincidental alphabetical cases. After Waltman [221].

Subject category	Alphabetical percentage
Biochemistry & Molecular Biology	−0.1%
Biology	0.2%
Medicine, General & Internal	0.2%
Materials Science, Multidisciplinary	0.6%
Neurosciences	0.8%
Chemistry, Multidisciplinary	0.9%
Chemistry, Physical	0.9%
Physics, Applied	1.3%
Engineering, Electrical & Electronic	3.6%
History	30.7%
Physics, Mathematical	32.1%
Statistics & Probability	33.8%
Mathematics, Applied	44.7%
Economics	57.4%
Physics, Particles & Fields	57.4%
Mathematics	72.8%

contributions, like providing the foundational idea for the project or correct proofs of key technical results needed by the authors. It is not uncommon for mathematicians to offer their colleagues authorship, only to be politely declined. So colleagues whose contributions were essential to the success of a paper end up merely acknowledged.

There are several benefits to the alphabetical ordering. For starters, it's convenient, allowing authors to avoid the awkward debate about whose name goes first or second. Furthermore, with the bar for authorship held high, alphabetization decreases the frequency of "guest" authors. (See Box 2.6.1.)

Box 13.2 Ghosts and guests

The trend of growing author lists has a troubling side effect: the rise of "guest authors," those who made only very minimal contributions to the research. A recent survey of 2,300 lead authors found that a shocking 33 percent of scholarly papers in the biological, physical or social sciences had at least one author whose contribution did not meet accepted definitions for coauthorship [166, 222]. The incidence of "undeserved" coauthors increases from 9 percent on papers with three authors to 30 percent on papers with more than six authors [213], with the most commonly cited reason for accepting undeserved authorship being academic promotion [141].

A far more serious concern are "ghost authors," individuals who made substantive contributions to a publication but who were not acknowledged as authors [223]. Take, for example, Robert Boyle, the most acclaimed chemist in seventeenth-century London. His laboratory was populated with assistants who tended distillations, amalgamations, and rectifications, made observations, and recorded them for Boyle's use [224]. Yet we know absolutely nothing about these researchers or their contributions to the research – not even their names – since Boyle published all their findings and contributions as his own (Fig. 13.3).

Unfortunately, this phenomenon has not abated in the centuries since. A recent estimate indicates that more than *half of the papers* in many disciplines have at least one ghost author [222]. The primary victims are often graduate students, whose relegation to ghost author status can hurt their chances of career advancement.

Figure 13.3 **Invisible technicians.** Artistic convention helped ensure the lowly technician's invisibility in the seventeenth century by depicting the agents operating scientific instruments as putti, or cherubs, rather than human beings. From Gaspar Schott, *Mechanica hydraulico-pneumatica* [225]. After Shapin [224].

Yet, alphabetical ranking also leaves out important information. Even if all authors have contributed equally, there is plenty of room, without explicit attribution, for the community to make (often inaccurate) inferences about who a paper's rightful "owner" is. As we will see next, this can have some serious consequences.

13.4 The Collaboration Penalty for Women

Economics, like many other disciplines, remains a stubbornly male-dominated profession. While women are disproportionately more likely to abandon the field at every stage, from high school to post-PhD [226–228], one especially leaky segment of the academic pipeline has puzzled economists for decades: Female economists are twice as likely to be denied tenure as their male colleagues [229, 230]. If we look at all faculty members in economics departments at the top 30 American universities, about 30 percent of them failed to win tenure at their first institution [230]. Yet, broken down by gender, a giant disparity emerges: forty-eight percent of women – nearly half of all female applicants! – were turned down for tenure, compared to just 23 percent of male applicants.

Why does such a chasm exist? Could it be rooted in different productivity patterns? The answer is no. If we measure the number of publications for both men and women economists before receiving tenure, and account for the prestige of the journals in which these papers were published, we find that the two groups are statistically indistinguishable [230]. Over the years researchers have investigated other plausible explanations, including heterogeneous tenure-granting rates across universities, which subfields of economics women work in, behavioral differences such as competitiveness and confidence [231], and the role of child-bearing. But it turns out, none of these factors can quite explain the gender gap in tenure outcomes [229].[1] As such, a significant portion (over 30%) of the gender promotion gap has remained a puzzling mystery [229, 230].

But recent research has found that the explanation for the gender gap in tenure rates may be quite simple [230]: Women

---

[1] To be sure, many of these variables do play a role. For example, the family commitment of female faculty is a significant factor, but it mostly explains why women take a year longer on average to be considered for tenure, and appears to have no impact on tenure outcome.

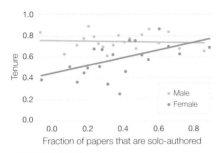

Figure 13.4 Team composition of papers and tenure rate. Correlation between tenure rate and the fraction of an individual's papers that are solo-authored, split by gender. Both variables are plotted after controlling for the number of years it took to go up for tenure, average journal rank for solo and coauthored publications, total citations, with tenure school, year, and field fixed effects. The line of best fit is shown separately for men and women ($N = 552$), which shows the slope equals to 0.41 for women and 0.05 for men. After Sarsons [230].

economists face an enormous penalty for collaborating (Fig. 13.4). That is, if we sort economists based on their tendency to collaborate – as measured by the fraction of solo-authored papers in an economist's career – female economists who always work alone have exactly the same success rate for receiving tenure as their male counterparts. Indeed, for both male and female economists, each solo-authored paper raises the probability of getting tenure by about 8 or 9 percent. Yet, as soon as their CVs start to include team-authored papers, the two genders part ways.

Men get just as much credit for collaborative research as for solo work, meaning that the career prospects of solitary economists and team players are effectively the same for men. However, when women write with female coauthors, the benefit to their career prospects is less than half that accorded to men. Things get especially dismal when women write with men. When a female economist collaborates with one or more male economists, her tenure prospects *don't improve at all*. In other words, when it comes to tenure, women get essentially zero credit if they collaborate with men. The effect is so dramatic that this single variable – the fraction of solo-authored papers – can explain most of gender differences in tenure outcomes.

Could this collaboration-based gender penalty be due to a lack of clear credit allocation protocols in economics? Sociologists, for example, collaborate almost as often, but they explicitly describe who deserves the most credit in a collaboration by listing that person as the first author. Perhaps as a result, tenure rates in sociology are comparable for men and women, even though men tend to produce more solo-authored papers than women. Hence, it appears that by listing authors in order of their contributions, it may help scientists eliminate the impulse to make inferences and assumptions based on bias. By contrast, most economics papers list authors alphabetically, a process that allows gender-based judgments, however unconscious, to have a devastating power.

*

In this chapter we compared science authorship protocols and their implications for scientific credit. However, the discussion so far only explores how credit *should* have been allocated in principle. But as the tenure gap for female economists illustrates, there are major issues with how scientific credit *is* allocated in practice. This is the focus of the next chapter.

14 CREDIT ALLOCATION

Even if you've never heard of the Thomas theorem, you're likely familiar with the idea. It states, in short: "If men define situations as real, they are real in their consequences." This helps explain, for example, how a mere rumor about a coming food shortage can make people flock to the grocery store to stock up, turning that rumor into a reality. This famous theorem, named after sociologist W. I. Thomas and often credited as the origin of the "self-fulfilling prophecy," first appeared in Thomas' book *The Child in America* [232], a deeply influential text in sociology. Both the Thomas theorem and *The Child in America* are repeatedly cited as the work of W. I. Thomas alone [233]. But the fact is, the cover of the book lists two authors: W. I. Thomas and Dorothy Swaine Thomas.

In this chapter, we will focus on the question of who ends up getting the credit for a collaborative work. We will see cases where certain researchers benefited, sometimes disproportionately, from a joint work, and other instances where they were unjustifiably overlooked. The truth is that how we allocate credit has a lot to do with bias, and some of these biases are deeply rooted in the way science is practiced. But the good news is, we now have a set of tools at our disposal that can decipher with increasing precision how the scientific community perceives the contribution of each author on any collaborative paper. These tools not only allow us to calculate how credit may be perceived and allocated for a given project – sometimes even before the collaboration begins – but also offer valuable insights about the intricacies and subtleties of how the scientific community allocates credit.

14.1 The Matthew Effect, Revisited

W. I. Thomas' unequal recognition for the joint work is an example of the Matthew effect, a process that governs how credit *is* (mis)allocated in science [234]. The Matthew effect states that, when scientists with different levels of eminence are involved in a joint work, the more renowned scientists get disproportionally greater credit, regardless of who did what in the project. Indeed, when Harriet Zuckerman interviewed Nobel laureates for her book [101], one physics laureate put it simply: "The world is peculiar in this matter of how it gives credit. It tends to give the credit to [already] famous people."

The most obvious explanations for this phenomenon are familiarity and visibility. That is, when we read through the list of authors on a

---

Box 14.1 The Matthew effect recap

We encountered the Matthew effect in Chapter 3, when we examined individual scientific careers. The Matthew effect operates through multiple channels in science which can be broadly categorized into (1) communication and (2) reward [234]. The Lord Rayleigh example pertains to the former, where scientific status and reputation influences the perception of quality. When his name was inadvertently omitted from the author list, the paper was promptly rejected. Yet, when Rayleigh's true identity was revealed, of course, the work itself remained the same, yet the response to the work changed.

In the reward systems of science, which is the focus of this chapter, the Matthew effect helps ensure that authors of vastly unequal reputation are recognized differently for their contributions. For example, the Matthew effect predicts that when independent discoveries are made by two scientists of different rank, the more famous one will get the credit. The story of Bernard of Chartres is a prime example. Although the twelfth-century philosopher coined the saying "Standing on the shoulders of giants" 400 years before Newton penned it in a letter to Robert Hooke, the famous phrase is nevertheless universally attributed to Newton. The Matthew effect also applies when multiple people are involved in the *same* discovery, but a disproportionate share of the credit ends up going to the most eminent collaborator. In this chapter, we treat both cases of the Matthew effect as stemming from the same credit attribution problem.

paper, the names we don't recognize are effectively meaningless to us, while a name with which we're familiar stands out prominently – and so we immediately associate the paper with the person we already know. More often than not, the familiar name is a more well-known and senior coauthor. In the words of a laureate in chemistry: "When people see my name on a paper, they are apt to remember *it* and not to remember the other names" [234]. Dorothy Swaine Thomas, whose contribution to *The Child in America* was diminished to the point of vanishing, represents this kind of credit discrimination. When the book was first published in 1928, W. I. Thomas, then 65, was the president of the American Sociological Society, a belated acknowledgement of his longstanding rank as dean of American Sociologists [234]. Dorothy Swaine Thomas, still in her twenties, was working as W. I. Thomas' assistant and was virtually unknown to the rest of the scientific community.

The disproportionate credit eminent scientists get when they collaborate with less well-known colleagues poses a dilemma for both junior and senior scientists, as succinctly summarized in Merton's interview with one of the Nobel laureates [60]:

> You have a student; should you put your name on that paper or not? You've contributed to it, but is it better that you shouldn't or should? There are two sides to it. If you don't, there's the possibility that the paper may go quite unrecognized. Nobody reads it. If you do, it might be recognized, but then the student doesn't get enough credit.

Indeed, research confirms that status signals eminence, which can not only increase the visibility of the work but also influence how the community perceives its quality [60, 61, 65]. Plus, through the process of working side by side with the best, a young researcher can also acquire all kinds of tacit knowledge he might not otherwise access, such as insights about a mentor's working habits, how she develops research questions, and how she responds to challenges along the way.

But from a credit perspective, there are serious downsides. Even if you had a chance to coauthor with Einstein, the paper would, first and foremost, always be Einstein's. Even though the junior author may have executed most of the work – which is frequently the case – their efforts will be overshadowed, and they'll end up getting less credit than they deserve.

And another important scenario to ponder: What happens when something goes wrong with the paper, such as a retraction, an issue with plagiarism, or when results are falsified or debunked? Who will be considered at fault? Eminent authors benefit more than they deserve from successful collaborations, precisely because the scientific community assumes that they were the source of the key ideas, while junior coauthor(s) are thought to have merely executed the project. Therefore, it is only reasonable to assume that the eminent author, as the presumed leader of the work, would bear more blame for errors or failures. With greater credit comes greater responsibility – right?

Research shows that exactly the opposite is true. In fact, when senior and junior authors are involved with the *same* retracted papers, senior authors can escape mostly unscathed from the fallout of the retraction while their junior collaborators (typically graduate students or postdoctoral fellows) are often penalized [75], sometimes to a career-ending degree. Therefore, a strong reputation not only assigns disproportionate credit to an eminent researcher; it also protects them in the event of a catastrophe. The underlying logic is quite simple: If Einstein coauthors a paper that is retracted, how could such a retraction possibly be Einstein's fault?

Box 14.2 The reverse Matthew effect, recap

As you may recall, when researchers compared retracted papers to closely-matched control papers [72] and non-retracted control authors [74], they found that retractions led authors to experience citation losses in their prior body of work [72–74], and eminent scientists were more harshly penalized than their less-distinguished peers in the wake of a retraction [74]. (See also Box 3.2 From boom to bust in Chapter 3.) Confusingly, these results appear to be at odds with what is discussed here. The key distinction between the two findings is whether we compare the citation penalties *within* teams or *between* them. When authors are on the same team, which is what this chapter concerns, the community assumes the fault lies with junior authors. Yet, if we compare two different papers, one by eminent authors and the other by relatively unknown authors, the results in Chapter 3 show that the former are more harshly penalized than the latter.

These results reveal a harsh reality about science: Credit is collectively determined by the scientific community, not by individual coauthors. Indeed, individual team members are often left helpless when their true contribution differs from what is perceived by the community. Still, there's a silver lining. Scientific credit follows precise rules that are not easily swayed by individual wills – which means, if we can decipher what these rules are, we can then follow them to calculate and even predict who will be seen as responsible for a discovery. By doing so, we can empower scientists to protect their share of the credit on team projects, and to ensure fair allocation of credit to coauthors.

14.2 Collective Credit Allocation

The paper shown in Fig. 14.1, which reported the discovery of the W and Z bosons [235], was awarded the Nobel Prize in Physics the

EXPERIMENTAL OBSERVATION OF ISOLATED LARGE TRANSVERSE ENERGY ELECTRONS WITH ASSOCIATED MISSING ENERGY AT $\sqrt{s}$ = 540 GeV

UA1 Collaboration, CERN, Geneva, Switzerland

G. ARNISON[j], A. ASTBURY[j], B. AUBERT[b], C. BACCI[i], G. BAUER[1], A. BÉZAGUET[d], R. BÖCK[d], T.J.V. BOWCOCK[f], M. CALVETTI[d], T. CARROLL[d], P. CATZ[b], P. CENNINI[d], S. CENTRO[d], F. CERADINI[d], S. CITTOLIN[d], D. CLINE[1], C. COCHET[k], J. COLAS[b], M. CORDEN[c], D. DALLMAN[d], M. DeBEER[k], M. DELLA NEGRA[b], M. DEMOULIN[d], M. DENEGRI[k], A. Di CIACCIO[i], D. DiBITONTO[d], L. DOBRZYNSKI[g], J.D. DOWELL[c], M. EDWARDS[c], K. EGGERT[a], E. EISENHANDLER[f], N. ELLIS[d], P. ERHARD[a], H. FAISSNER[a], G. FONTAINE[g], R. FREY[h], R. FRÜHWIRTH[l], J. GARVEY[c], S. GEER[g], C. GHESQUIÈRE[g], P. GHEZ[b], K.L. GIBONI[a], W.R. GIBSON[f], Y. GIRAUD-HÉRAUD[g], A. GIVERNAUD[k], A. GONIDEC[b], G. GRAYER[j], P. GUTIERREZ[h], T. HANSL-KOZANECKA[a], W.J. HAYNES[j], L.O. HERTZBERGER[2], C. HODGES[h], D. HOFFMANN[a], H. HOFFMANN[d], D.J. HOLTHUIZEN[2], R.J. HOMER[c], A. HONMA[f], W. JANK[d], G. JORAT[d], P.I.P. KALMUS[f], V. KARIMÄKI[e], R. KEELER[f], I. KENYON[c], A. KERNAN[h], R. KINNUNEN[e], H. KOWALSKI[d], W. KOZANECKI[h], D. KRYN[d], F. LACAVA[d], J.-P. LAUGIER[k], J.-P. LEES[b], H. LEHMANN[a], K. LEUCHS[a], A. LÉVÊQUE[k], D. LINGLIN[b], E. LOCCI[k], M. LORET[k], J.-J. MALOSSE[k], T. MARKIEWICZ[d], G. MAURIN[d], T. McMAHON[c], J.-P. MENDIBURU[g], M.-N. MINARD[b], M. MORICCA[i], H. MUIRHEAD[d], F. MULLER[d], A.K. NANDI[j], L. NAUMANN[d], A. NORTON[d], A. ORKIN-LECOURTOIS[g], L. PAOLUZI[i], G. PETRUCCI[d], G. PIANO MORTARI[i], M. PIMIÄ[e], A. PLACCI[d], E. RADERMACHER[a], J. RANSDELL[h], H. REITHLER[a], J.-P. REVOL[d], J. RICH[k], M. RIJSSENBEEK[d], C. ROBERTS[j], J. ROHLF[d], P. ROSSI[d], C. RUBBIA[d], B. SADOULET[d], G. SAJOT[g], G. SALVI[f], G. SALVINI[i], J. SASS[k], J. SAUDRAIX[k], A. SAVOY-NAVARRO[k], D. SCHINZEL[f], W. SCOTT[j], T.P. SHAH[j], M. SPIRO[k], J. STRAUSS[1], K. SUMOROK[c], F. SZONCSO[1], D. SMITH[h], C. TAO[d], G. THOMPSON[f], J. TIMMER[d], E. TSCHESLOG[a], J. TUOMINIEMI[e], S. Van der MEER[d], J.-P. VIALLE[d], J. VRANA[g], V. VUILLEMIN[d], H.D. WAHL[1], P. WATKINS[c], J. WILSON[c], Y.G. XIE[d], M. YVERT[b] and E. ZURFLUH[d]

Aachen [a] – Annecy (LAPP) [b] – Birmingham [c] – CERN [d] – Helsinki [e] – Queen Mary College, London [f] – Paris (Coll. de France) [g] – Riverside [h] – Rome [i] – Rutherford Appleton Lab. [j] – Saclay (CEN) [k] – Vienna [l] Collaboration

Received 23 January 1983

Figure 14.1 The Nobel Prize in Physics 1984. Screenshot of the paper reporting the discovery of the particles W and Z [235], based on which the 1984 Nobel Prize in Physics was awarded jointly to two of the paper's coauthors, Carlo Rubbia and Simon van der Meer, "for their decisive contributions to the large project, which led to the discovery of the field particles W and Z, communicators of weak interaction."

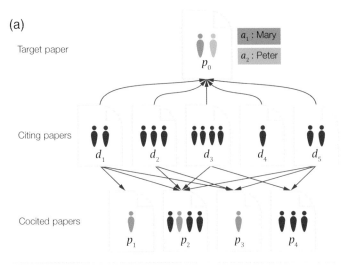

(b) 2010 Nobel Prize in Chemistry
Baba, Negishi, J. Am. Chem Soc. 98, 6729 (1976).

Cocited Papers
Negishi, Okukado, King, Van Horn, Spiegel, J. Am. Chem. Soc. (1978)
Negishi, King, Okukado, J. Org. Chem. (1977)
Negishi, Vanhorn, J. Am. Chem. Soc. (1977)
Negishi, Vanhorn, J. Am. Chem. Soc. (1978)
Negishi, Valente, Kobayashi, J. Am. Chem. Soc. (1980)

(c) 2010 Nobel Prize in Physics
Novoselov, Geim, Science, 306, 666 (2004).

Cocited Papers
Geim, Novoselov, Nature (2007)
Novoselov, Jiang, Schedin, Booth, Khotkevich, Morozov, Geim, PNAS (2005)
Novoselov, Geim, Morozov, Jiang, Katsnelson, Grigorieva, Dubonos, Firsov, Nature (2005)
Castro Neto, Guinea, Peres, Novoselov, Geim, Rev. Mod. Phys. (2009)
Ferrari, Meyer, Scardaci, Casiraghi, Lazzeri, Mauri, Piscanec, Jiang, Novoselov, Roth, Geim, Phys. Rev. Lett. (2006)

Figure 14.2 **Collective credit allocation in science.** (a) Illustration of one extreme case of how credit allocation works in science: When Peter contributes to only one paper in a body of work, the community assigns most of the credit to Mary, who publishes multiple papers on the topic. (b) and (c) show two case studies of credit allocations. Negishi exemplifies the example shown in (a), as he published several other papers in this area that are cocited by the community with the Nobel-winning paper on the top. Yet, the prizewinning paper was the only one published by his coauthor Baba. By contrast, in (c), Novoselov and Geim were jointly involved in almost all highly cocited papers on the subject and were jointly awarded the Nobel Prize with an equal share. After Shen and Barabási [236].

year after its publication, an exceptionally quick turn-around between discovery and recognition. But who of the 135 authors alphabetically listed on the paper deserved the Nobel? For high-energy physicists, the answer is a no-brainer: The credit undoubtedly goes to Carlo Rubbia and Simon van der Meer – and only to the two of them – given their "decisive contributions to the large project," to use the words of the Nobel committee. Yet, for anyone else, picking these two out of a long list of authors listed in alphabetical order seems like nothing short of magic.

Clearly, the scientific community uses an informal credit allocation system that requires significant domain expertise. But for those outside of the discipline, is there a way to know who was central to a discovery? In Chapter 13, we learned some useful rules of thumb to infer credit. For example, when authors are ordered based on their contributions, we know that we should often pay attention to the first and last authors. But, if you ask us to pick two out of over a hundred names listed alphabetically as in Fig. 14.1 – we are lost. Thankfully, we have developed an algorithm that performs this magic, capturing the collective process through which scientific credit is allocated [236], and offering a tool for calculating the share of credit among coauthors for any publication.

To understand how the algorithm works, imagine two authors, Mary and Peter, and their joint publication p_o (Fig. 14.2a). Who gets the credit for this paper? One way to surmise the answer is by examining who has more proven expertise on the subject matter. For instance, consider one extreme case where Mary has published several other papers on the topic, whereas p_o is Peter's only publication on this topic. This fact can be detected through citation patterns: if we look at papers that tend to be cited together with p_o, several of them will be authored by Mary, yet none by Peter. This means Mary has a well-established reputation in the community, whereas Peter is effectively an outsider. As a result, the community is more likely to view the paper p_o, as a part of Mary's body of work, rather than Peter's (Fig. 14.2a).

Ei-ichi Negishi, who won the 2010 Nobel Prize in Chemistry for his joint 1976 paper with Shigeru Baba (Fig. 14.2b) exemplifies this scenario. Although Baba coauthored the landmark paper, it was his only high-profile publication on the topic. In contrast, Negishi published several other highly cited papers in subsequent years. When the Nobel committee had to decide, they awarded the prize to Negishi alone.

Consider the other extreme case, shown in Fig. 14.2c. In this case, *all* highly cited papers pertaining to the topic of p_o are joint publications between the same two scientists, Novoselov and Geim. If the two scientists always share authorship, an observer without any outside information must give both authors equal credit for p_o. Their joint Nobel Prize for the discovery of graphene is an excellent example of this scenario: Novoselov and Geim not only coauthored the first paper on graphene, but were also jointly involved in almost all high-impact publications on the topic afterwards, earning them an equal share of credit for the original discovery.

Clearly, however, most cases are more complicated, lying somewhere in between the two extremes. The two authors may publish some papers together, and some papers with other coauthors on similar topics. Hence their credit share for a particular work may change with time. So how can we quantify the proportion of "the body of work" allocated to each coauthor? We can inspect the cocitation patterns between the paper in question and all other papers published by the same authors. Figure 14.3 illustrates a collective credit allocation algorithm that allows us to do just that [236]. To get a sense of how this works, we discuss credit allocation for a paper with two arbitrary coauthors in Fig. 14.3.

In the algorithm outlined in Fig 14.3, each additional paper conveys implicit information about the author's perceived contribution, capturing, in a discipline-independent fashion, how the scientific community assigns credit to each paper. This algorithm is not only a theoretical tool but a predictive tool: when applied to all multi-author Nobel prize-winning publications, the method correctly identified the laureates as the authors deserving the most credit 81 percent of the time (or, in 51 out of 63 papers). Regardless of whether the winner was in high physics, where authors are listed alphabetically, or in biology, where a team's leader is usually the first or last author, the algorithm consistently predicted the winners. And in the 19 percent of cases where the algorithm picked the wrong person, it often revealed interesting tensions and potential misallocations worthy of further consideration (see Box 14.3).

The credit allocation algorithm implies that when we allocate credit to team members, we don't need to know who actually did the work. This underlines an important message: Scientific credit is not just about contribution – it's also about perception. As such, this chapter

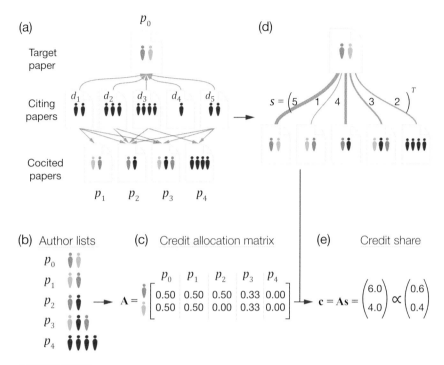

Figure 14.3 An illustration of a collective credit allocation algorithm [236]. (a) The target paper p_o has two authors, colored in red and green, respectively. To determine the credit share going to each of the two authors for their joint paper p_o, we first identify all papers that cite p_o: $\{d_1, \ldots, d_5\}$. We then trace their references to identify all other papers cocited with p_o, $P \equiv \{p_o, p_1, \ldots, p_4\}$. The set of papers P thus represent the total body of work specifically related to p_o, as every time p_o is referenced in the literature, it appeared together with papers in P. Next, we seek to discover the involvement of the two authors within this body of work P. As shown in (b), we can first go through all papers in P and calculate their involvement in p_o, $p_1, \ldots, p_4$. In the example shown in (a), both authors are assigned equal (half) credit for p_o and p_1, as they are the only two authors involved in these papers. But for p_2, only the red author is involved within a two-author team, hence they are assigned 0.5 credit, while the green author receives 0 for this paper. Repeating this process for all papers in P, we can obtain an authorship involvement matrix $\mathbf{A}$, where each element A_{ij} denotes the amount of credit that each author gets from each paper p_j published on this topic. (c) The credit allocation matrix $\mathbf{A}$ obtained from the author lists of the cocited papers in (b). The matrix $\mathbf{A}$ provides the author's share for each cocited paper. For example, because p_2 has the red author as one of its two authors but it lacks the green author, it provides 0.5 for the red author and 0.0 for the green author. Yet, not all papers in P are the same: some papers are more relevant than others to p_o. This can be resolved by calculating the co-citation strength s_j between p_o and any given paper in P, p_j, which measures the number of times p_o and p_j are cited together. This process is illustrated in (d), showing the

doesn't just offer us a quantitative tool that can help us calculate the perceived credit share for joint publications. It also offers several insights regarding the nature of collaboration.

Lesson 1: In science, it is not enough to publish a breakthrough work and simply wait to be recognized for it. To claim your well-deserved credit in the eyes of others, you must continue publishing work of importance independent of your previous coauthors. You need to be seen by the community as the consistent intellectual leader behind your earlier breakthrough work, and your contribution to the original paper matters little in shaping this perception. Rather, as the algorithm illustrates, what you publish *after* the breakthrough in question is what ultimately determines your share of credit.

Lesson 2: If you venture into an established area – for instance, beginning your career working with experts in your field – your credit share may be pre-determined at the outset. Indeed, if your coauthors have done significant work on a topic you are just beginning to explore, their previous work will likely be cocited with the current paper. In this case, even if you subsequently manage to publish many relevant works afterwards – and do so independently of your original coauthors – it may take a long time for you to overcome the pre-existing credit deficit, if it ever happens at all. After all, if papers that cite your new work also cite past canonical works by your coauthors in the area, it will dilute your additional effort.

There are many reasons to pursue a particular research project. We may join a team because we like working with the team members, because we believe in its mission, or because there is a big problem to be

---

Figure 14.3 (cont'd) p_0-centric cocitation network constructed from (a), where the weights of links denote the cocitation strength s between the cocited papers and the target paper p_0. For example, in (a), p_1 has only been cited once with p_0 (through d_1), whereas p_2 and p_0 have been cited in tandem by four different papers (d_1, d_2, d_3, d_5). Therefore, insofar as p_0 is concerned, p_2 should have a higher cocitation strength (i.e., $s_1 = 1$ vs. $s_2 = 4$) than p_1, meaning it is likely to be more closely related. (e) With the matrix A and cocitation strength s, the credit share of the two authors of p_0 is computed by the multiplication of the two matrices with a proper normalization. The ultimate credit allocated to each author by the algorithm is simply their involvement in the body of work, weighted by the relevance of that work. Mathematically, this corresponds to the multiplication of author involvement matrix and cocitation strength matrix, $c = As$. After Shen and Barabási [236].

Box 14.3 When the credit allocation algorithm fails

While it's fascinating to see the algorithm automatically identify prizewinners from long lists of names, it's equally interesting to witness the method fail [236]. This would imply that there are individuals who, in the eyes of the community, likely deserve major credit for a discovery, but were nevertheless overlooked by the Nobel committee. One example is the 2013 Nobel Prize in Physics, awarded for the discovery of the "God particle," the Higgs boson. Six physicists are credited for the 1964 theory that predicted the Higgs boson, but the prize could only be shared by a maximum of three individuals. F. Englert and R. Brout published the theory first [237], but failed to flesh out the Higgs boson, whose existence was predicted in a subsequent paper by P. W. Higgs [238]. G. S. Guralnik, C. R. Hagen, and T. W. B. Kibble independently proposed the same theory and published it a month after Englert and Brout [239], explaining how the building blocks of the Universe get their mass. In 2010, the six physicists were given equal recognition by the American Physical Society (APS), sharing the Sakurai prize for theoretical particle physics. Yet, the Nobel committee awarded the prize only to Higgs and Englert in 2013. The credit allocation algorithm predicts that Higgs would get the most credit, followed by Kibble. Englert, however, is the third, with only a slightly higher credit share than his coauthor Brout, who died before the Noble was awarded. Guralnik and Hagen equally share the remaining credit. This means that, according to the algorithm, the 2013 Nobel Prize for Physics should have gone to Higgs, Kibble, and Englert in that order. By passing over Kibble, the committee deviated from the community's perception of where the credit lies.

A similar mismatch between the real-world laureates and the algorithm's prediction brought Douglas Prasher to our attention. In 2008, when the Nobel prize was awarded, his credit share for the discovery of GFP was 0.199, thanks to the many highly cited key papers he coauthored on the topic. While this share was smaller than that of two of the laureates, Tsien and Shimomura (0.47 and 0.25, respectively), it exceeded the share of Martin Chalfie (0.09), who also received the award.

solved. In other words, we often join a team without caring about who gets the credit for the work. But if we *do* seek credit, then the algorithm can help us estimate ahead of time whether we can ever hope to gain a noticeable credit for the team's work (see also Box 14.4).

> **Box 14.4 The credit allocation algorithm may raise incentive issues**
>
> By quantitatively exposing the informal credit system, algorithms, like the one we discussed in this chapter, allow anyone to calculate their perceived credit share even before a team has officially formed, which may have unintended consequences [105, 147]. For example, individuals may be encouraged to select collaborators partly based on ex-post credit considerations rather than the effectiveness of the team itself. Will the availability of such methods make collaborations more strategic? And if so, what can we do to counterbalance personal interests against the collective advance of science? We do not have answers to these questions. But, as our approach to understanding the mechanisms within science becomes increasingly more scientific, it would allow us to anticipate and detect new side effects earlier, and with a better diagnosis.

*

Alas, Dorothy Swaine Thomas faced two challenges at the time – she was not only a junior researcher collaborating with an established partner, but also a woman in an era when the scientific community regularly downplayed the contributions of women. However, her story does not end in obscurity: She eventually went on to have a distinguished scientific career, achieving a prominent reputation of her own. Like W. I. Thomas, she too was elected to the presidency of the American Sociological Society in 1952. But none of these achievements prevented even the most meticulous scholars from attributing the 1928 book solely to her famous coauthor, a practice that continues even today. What's more, as this chapter makes clear, had it not been for her later stellar career, we may not be grumbling about the omission of her name in the first place – she might have been lost to history entirely.

Part III THE SCIENCE OF IMPACT

"If I have seen further, it is by standing on the shoulders of giants," wrote Isaac Newton in a letter to Robert Hooke on February 1676 [240]. That oft-quoted line succinctly captures a fundamental feature of science: its cumulative nature. Indeed, scientific discoveries rarely emerge in isolation but tend to build on previous work by other scientists. Scientists throughout the ages have typically acknowledged the provenance of the ideas they built upon. Over time this custom has turned into a strict norm, giving rise to *citations*.

A citation is a formal reference to some earlier research or discovery, explicitly identifying a particular research paper, book, report, or other form of scholarship in a way that allows the reader to find the source. Citations are critical to scientific communication, allowing readers to determine whether previous work truly supports the author's argument. Used appropriately, they help the author bolster the strength and validity of their evidence, attribute prior ideas to the appropriate sources, uphold intellectual honesty, and avoid plagiarism. For the author, citations offer a vehicle to condense and signal knowledge and eliminate the need to reproduce all of the background information necessary to understand the new work.

Lately, however, citations have taken on an additional role: the scientific community has begun using them to gauge the scientific impact of a particular paper or body of work. This is possible, the thinking goes, because scientists only cite papers that are relevant to their work. Consequently, groundbreaking papers that inspire many scientists and research projects should receive many citations. On the other hand, incremental results should receive few or no citations.

Yet as straightforward as this linkage between impact and citations may appear, it is fraught with ambiguity: How many citations constitutes "a lot"? What mechanisms drive the accumulation of citations? And what kinds of discoveries tend to garner more citations? Can we know how many citations a paper will collect in the future? How soon can we tell if a discovery has sparked serious interest? We must also take a step back and ask what citation counts tell us about the validity and impact of the ideas described in the paper. Put simply: Are citations meaningful at all?

We seek quantitative answers to these questions in the next six chapters. We have previously examined a range of patterns underlying individual scientific careers and collaborations in the first two parts of this book. Now that we have understood a lot more about the "producers" of science, be it individuals or teams, it is now time to focus on what they produce. Our first question is therefore, how many papers have we produced? For that, we need to go back to 1949 and pay a visit to Singapore.

15 BIG SCIENCE

In 1949, Derek de Solla Price was teaching applied mathematics at Raffles College in Singapore when the college's new library received a complete set of the *Philosophical Transactions of the Royal Society of London*. Published since 1662, *Philosophical Transactions* was the first journal exclusively devoted to science. As the library was still waiting for its future building, de Solla Price kept the volumes at his bedside. He spent the next year reading the journals cover to cover, sorting each decade of volumes into a separate pile. One day, as he looked up from his reading, he made a curious observation: the pile for the first decade was tiny, and the second only slightly taller. But as the decades went on, the height of the piles began to grow faster and faster, accelerating wildly in the most recent decades. Altogether, the 28 piles formed what looked like a classic exponential curve.

It was an observation that sparked a lifetime of passion – in the following decades, de Solla Price would systematically explore how science grew over time. By 1961, he had counted everything he could get his hands on: the number of scientific journals, the number of scientists who contributed to them, and the total number of abstracts in multiple fields [241]. No matter which dimension he charted, the same exponential curve greeted him, confirming his original insight that science was accelerating [241]. "[E]xponential growth, at an amazingly fast rate, was apparently universal and remarkably long-lived," he concluded [12].

But, de Solla Price was quick to realize, such exponential growth is rarely sustainable. We know that from studies on bacterial colonies: the number of bacteria grows exponentially in the beginning, but eventually the nutrients that sustain their growth are exhausted, and the growth saturates. Hence, de Solla Price predicted that the exponential growth of science must represent merely an initial growth spurt. Science must run out of steam and saturate eventually. He even went as far as to predict that the rapid growth in science should taper off shortly after the 1950s [242].

As we show in this chapter, while his observation was real, his prediction was wrong. The exponential growth of scientific research has continued virtually uninterrupted since de Solla Price's time, dramatically reshaping the world along the way.

15.1 The Exponential Growth of Science

Figure 15.1 shows the number of publications indexed yearly by Web of Science (WoS), documenting that for over a century the number of published papers has been increasing exponentially. On average, the total number has doubled roughly every 12 years. The

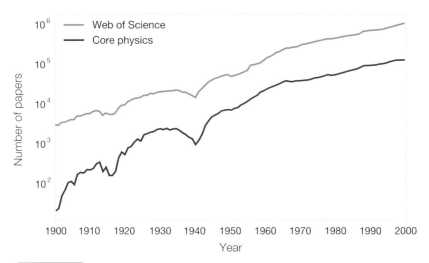

Figure 15.1 **The growth of science.** The number of papers catalogued in the Web of Science (WoS) published over the past century, illustrates the exponential growth of the scientific literature. It was only disrupted around 1915 and 1940 due to the World Wars. The figure also shows the growth of the physics literature, which follows an exponential growth similar to that followed by science as a whole. After Sinatra et al. [116].

figure also shows that science did not saturate after the 1950s as de Solla Price expected. Rather, its exponential growth has continued for the past 110 years, an expansion that was halted only temporarily by the two World Wars. Moreover, if we measure the growth rate of a single discipline, such as physics, we find similar exponential growth (Fig. 15.1). The fact that individual disciplines have accelerated in the same fashion suggests that the expansion of science is not simply driven by the emergence of new fields, but is instead a feature that applies across all areas of science.

What has the exponential growth of science meant for science, and for scientists? In this chapter, we will attempt to answer these complex questions.

Box 15.1 How many papers are there?

Considering the exponential growth of science over the centuries, you might wonder how many papers science has produced. In 2014, researchers estimated the total number of scholarly documents on the Internet using a "mark and recapture" method [243] (originally developed by ecologists to estimate the population size of wild animals when they cannot count every individual). They found that back then, there were at least 114 million English-language scholarly documents accessible on the Web. Obviously, given the size of the Internet, we will never know the precise number of scholarly documents out there. But whatever the estimated number may be at any given point in time, it would quickly become outdated. Indeed, thanks to the exponential growth of science, by March 2018, four years after the 114 million estimated, Microsoft Academic Search has indexed more than 171 million papers [244].

15.2 The Meaning of Exponential Growth

Any system following an exponential growth must expand at a rate proportional to its current size. Since the scientific literature doubles roughly every 12 years, this means that of all scientific work ever produced, half of it has been produced in the last 12 years. Thus, science is characterized by immediacy [12]: the bulk of knowledge remains always at the cutting edge. Furthermore, with an ever-growing number of scientists, scientists are often contemporaneous with those who have revolutionized their discipline. This is well captured by de Solla Price in his book, *Little Science, Big Science, and Beyond* [12]:

> During a meeting at which a number of great physicists were to give firsthand accounts of their epoch-making discoveries, the chairman opened the proceedings with the remark: "Today we are privileged to sit side-by-side with the giants on whose shoulders we stand." This, in a nutshell, exemplifies the peculiar immediacy of science, the recognition that so large a proportion of everything scientific that has ever occurred is happening now, within living memory. To put it another way, using any reasonable definition of a scientist, we can say that 80 to 90 percent of all the scientists that have ever lived are alive now.

To understand what this exponential growth has meant for individuals, imagine a young scientist at the beginning of her career [12]. After years of reading the literature, with guidance from knowledgeable mentors, she has finally reached the frontier of her discipline, and is ready to strike out on her own. If science had stopped growing years ago, she would be voyaging alone into uncharted seas. While happily making discovery after discovery, she would also be quite lonely, having relatively few peers with whom to collaborate or learn from. Hence, in this respect, the continual growth of science is good news, supplying ample like-minded colleagues and interesting ideas, so that scientists can build on the work of one another and explore the world of unknowns together.

But the rapid expansion of science gives rise to another problem. As our young scientist unties the rope and tries to sail away from the harbor, she will find that there are many other boats, big and small, headed in the same direction, captained by peers with the same level of training, ambition, determination, and resources. This competitive environment will have serious consequences for our young scientist throughout her career.

First, try as she might, it will not be possible for her to monitor closely where each boat is headed. While the volume of new knowledge grows exponentially, the time a scientist can devote to absorbing new knowledge remains finite. As such, it is impossible for an individual today to read every paper in her own field.

Perhaps more importantly, each of those boats is hoping to discover something new. Indeed, in science, priority for a discovery has always been of key importance. Our young scientist would rather chart a new water than to sail in the wake of others. But so much competition may affect the individual's odds of making that grand discovery.

Many people have a heroic conception of science, believing that the field is moved forward by a small number of geniuses. But in reality, groundbreaking discoveries are often the culmination of years of hard work by many scientists. A "Eureka" moment happens when someone acknowledges what came before and manages to take the next logical leap. Indeed, Francis Bacon once argued, all innovations, social or scientific are "a birth of time rather than of wit" [245]. In other words, apples fall when they are ripe, regardless of who is around to harvest them. Thus, if a scientist misses a particular discovery, someone else will discover it instead. After all, if we already have a steam engine and a boat, how long will it take for someone to invent a steamboat?

This perhaps underlies one of the most important ways in which science differs from other creative endeavors. If Michelangelo or Picasso had never existed, the sculptures and paintings we admire in art museums would be quite different. Similarly, without Beethoven, we would not have the 5th Symphony, nor the distinctive "Pa pa pa PAM!" However, if Copernicus had never existed, we would not have arrived at an alternative description of the solar system – sooner or later we would have figured out that the Earth orbits around the Sun, and not the other way around.

This has important implications for scientists: Apples may not care who harvests them – but for apple pickers, being the first to reach a ripe apple is critical. If science is a race to be the first to report a discovery, then the exponential growth of science poses a major question: Did the increasing competition make it more difficult to practice science?

The answer is not obvious. After all, as spectacular as the recent growth of science has been, the historical record suggests that pressures to digest and create new knowledge are not new. For example, in 1900, after being scooped by Pierre and Marie Curie on a radioactivity paper, Ernest Rutherford wrote: "I have to publish my present work as rapidly as possible in order to keep in the race" [246] (see Box 15.1). Or consider this confession [12]:

> One of the diseases of this age is the multiplicity of books; they doth so overcharge the world that it is not able to digest the abundance of idle matter that is every day hatched and brought forth into the world.

This passage by Barnaby Rich dates back to 1613, half a century before the publication of the first scientific journal. Clearly the experience of feeling overwhelmed by the quantity of existing knowledge – and daunted by the prospect of adding to it – predates our generation. So, is practicing science really getting harder? To find out, let's break down the steps required to become a successful scientist, examining how each step has evolved in this age of big science.

Box 15.2 The birth of "letters"

Letters are a common publication format in many journals, such as *Nature* or *Physical Review Letters*. But the title may seem deceiving; these "letters" are usually complete research papers, rather than anything resembling a letter. So where does the name come from? It traces back to Ernest Rutherford, whose pragmatic choice of communicating his research shaped the way we disseminate scientific knowledge today [247].

At the beginning of the twentieth century, being the first to make a discovery was already an important goal; however, very few journals published frequently enough to allow for the rapid dissemination of new discoveries [248]. One such journal was *Nature*, which was published weekly and was read by European (mainly British) scientists, Rutherford's primary competition and audience. He cleverly started using the "Letters to the Editor" section – a column traditionally dedicated to comments on someone else's work – as a way to rapidly communicate his own discoveries and establish priority. If you read Rutherford's "letters" as far back as 1902, you will find that they look more like scientific papers than short commentaries. This Trojan horse approach proved so successful that it was soon adopted by Rutherford's students and collaborators, notably Otto Hahn, Niels Bohr, and James Chadwick. By the 1930s Otto Frisch, Lise Meitner, and the "Copenhagen school" of physicists working with Bohr had adopted it, eventually morphing the letters section into the format we know today.

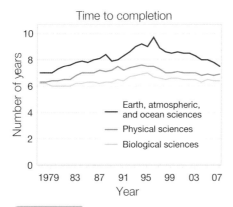

Figure 15.2 How long does it take to get a PhD in the United States? In 2007, it took a median of 7.2 years to complete a science or engineering PhD. After Cyranoski et al. [252].

15.3 Is it Getting Harder to Become a Scientist?

Becoming a scientist these days requires a number of essential steps, from getting the appropriate degree, to finding employment that allows the individual to pursue science. The rapid growth of training opportunities and that of the scientific workforce have affected these steps.

15.3.1 Obtaining a PhD

Figure 15.2 plots the time it takes to complete a science PhD in the United States, showing that the process today takes only slightly longer than it did 40 years ago. Despite a slight downward trend in the last two decades, the average time to degree in the life sciences and engineering remains six to eight years. In fact, the five years quoted by most PhD programs, remains a dream for most candidates – less than 25 percent of students manage to complete degrees within this period [249, 250]. Instead, 55 percent of candidates require seven or more years to complete their degree. And these statistics only count the individuals who do obtain a degree, obscuring the fact that 40 percent to 50 percent of candidates who begin their doctoral education in the US never graduate [249, 251].

It certainly seems easier than ever to begin doctorate training, given the rapidly growing number of programs. But, why does it take so long to acquire a PhD? Multiple measures show that, once enrolled in the program, it is getting harder to produce the dissertation required for graduation.

Consider, for example, the length of the thesis itself: The average length of biology, chemistry, and physics PhD theses nearly doubled between 1950 and 1990, soaring from 100 pages to nearly 200 pages in just four decades [76]. The number of references contained in a paper has been also rising over the years [253], indicating that research papers today build on a larger body of previous knowledge than ever before. Finally, the bar for publishing that dissertation – often a key milestone for aspiring scientists – has also grown higher: Comparing biology papers published in three major journals in the first six months of 1984 with the same period in 2014, researchers found that the number of panels in experimental figures (the charts and graphs published alongside papers) rose two- to four-fold [254], indicating that the amount of evidence required for a successful publication has increased significantly.

These statistics make clear how hard it has become to earn a PhD. But completing one's degree is only the first step toward a successful career in science.

15.3.2 Securing an Academic Job

Price's observation that "80 percent to 90 percent of the scientists who ever lived are alive now" memorably illuminates the exponential growth of the scientific workforce. Yet not all parts of the academic pipeline have expanded equally quickly. For example, the number of science doctorates earned each year grew by nearly 40 percent between 1998 and 2008, reaching 34,000 across the 34 countries that constitute the Organization for Economic Co-operation and Development (OECD) [252]. In the same three decades, however, the number of faculty positions in those nations has remained largely unchanged and even fallen slightly (Fig. 15.3) [251].

While not everyone with a PhD aims to secure an academic position, the vast majority of PhDs today do seem to prefer an academic career to alternatives in industry, government, or non-profits. In a world-wide survey conducted by *Nature* in 2017, nearly 75 percent of the 5,700 doctoral students polled preferred a job in academia to these non-academic alternatives [255]. For them, the trend shown in Fig. 15.3 may seem disheartening, showing a steadily growing PhD pool competing for a fixed number of opportunities.

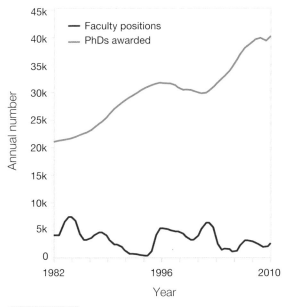

Figure 15.3 The academic pipeline. Since 1982, almost 800,000 PhDs have been awarded in science and engineering (S&E). These graduates competed for about 100,000 academic faculty positions. The number of S&E PhDs awarded annually gradually increased over this time frame, from 19,000 in 1982 to 36,000 in 2011. Yet the number of faculty positions created each year has remained stable or even fallen slightly. In 2010, the 40,000 PhD candidates competed for roughly 3,000 new academic positions. After Schillebeeckx et al. [251].

One consequence of increasing PhD production is a booming number of postdoctoral fellowships, representing short-term, non-tenure track research positions. As Fig. 15.5 illustrates, the number of researchers in US postdoctoral positions has more than tripled since 1979, with growth particularly prominent in the life sciences. Although postdocs drive research in many disciplines, they are often poorly compensated. Five years after receiving a PhD, the median salaries of scientists in tenure-track positions or industry far outstrip those of postdocs.

Such bleak prospects for landing a full-time university job means that many scientists have to seek alternatives outside academia. But how well do they fare once they leave science? To answer this question, researchers collected data in eight American universities and linked anonymized census data on employment and income to administrative records of graduate students at these universities [257]. Focusing

Box 15.3 Academic pipelines around the world (Figure 15.4)

The **United States** produced an estimated 19,733 PhDs in the life and physical sciences in 2009 [252], a number that has continued to increase ever since. But the proportion of science PhDs who secure tenure-track academic positions has been dropping steadily, and industry has not been able to fully absorb the slack. In 1973, 55 percent of US doctorates in the biological sciences had secured tenure-track positions six years after completing their PhDs, and only 2 percent were still in a postdoc or other untenured academic position. By 2006, the fraction of graduates who had secured a tenure-track position within six years dropped to 15 percent – but 18 percent were now holding untenured positions.

Figure 15.4 Academic pipelines around the world.

The number of PhD holders in **mainland China** is also going through the roof, with some 50,000 students graduating with doctorates across all disciplines in 2009, more than twice the number of US degree recipients. However, thanks to China's booming economy, most Chinese PhD holders quickly find jobs in the private sector.

Japan, too, has sought to get its science capacity on par with the West. In 1996, the country's government announced an ambitious initiative to increase the number of Japanese postdocs to 10,000. Yet, the government gave little thought to where all those postdocs would find jobs. Even though the government once offered companies a subsidy of around 4 million yen (US$47,000) to hire these well-educated individuals, the country still has 18,000 former postdocs who are still unemployed.

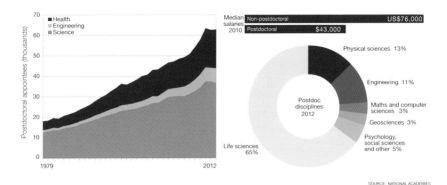

Figure 15.5 **The postdoc pileup.** The number of researchers in US postdoctoral positions has more than tripled since 1979. The vast majority of postdocs are in the life sciences. Across fields, median postdoc salaries are outstripped by those who choose non-postdoc careers, when measured up to five years after receiving a PhD. After Powell [256].

on those PhD students who graduated between 2009 and 2011, they found that almost 40 percent had entered industry, with many getting jobs at large, high-wage establishments in the technology and professional services industries. Top employers included electronics, engineering and pharmaceutical companies. These results show that PhDs who leave science tend to navigate toward large firms and earn a median salary of more than US$90,000. Related research also suggests that scientists in the private sector experience very low unemployment rates overall [258]. According to the National Science Foundation's Survey of Doctorate Recipients, just 2.1 percent of individuals with doctorates in science, engineering, or health in the United States were unemployed in 2013, while the overall national unemployment rate for people aged 25 and older was 6.3 percent.

Together, these data show that for the voyagers, the ocean of knowledge and opportunities *is* getting increasingly crowded. As the number of PhDs grows and the number of available university posts remains constant, competition intensifies, leading to a soaring number of postdocs, and more people leaving academia to pursue careers elsewhere. But it is important to note that this trend is not necessarily detrimental to society, nor to the scientists who pursue a wide range of careers. All signs so far indicate that PhDs who leave science do tend to find fulfilling jobs. And while they may not be creating knowledge in the scholarly sense, their work often leads to critical patents, products, and innovative solutions. As such, the loss for science is often a gain for society.

15.4   Is Science Running Out of Gas?

As we stated earlier, apples fall when they are ripe – that is, scientific breakthroughs are possible once a critical mass of knowledge has been amassed. But with more and more people trying to harvest ripe apples, has the remaining fruit on the tree become harder to reach? More specifically, do new discoveries require more effort now than they did in the past?

Tracking changes in productivity can help us answer this question. Economists define productivity as the amount of work-hours required to produce an output, like building a car or printing a book. In science, it corresponds to making a new discovery, measuring the amount of effort required to write a research paper. As we have shown, both scientific discoveries and the scientific workforce have been growing exponentially for decades. But depending on which has grown faster, the average productivity per scientist may be either rising or falling. And if scientific productivity has indeed decreased, that would imply that more labor is now required to achieve any breakthrough – in other words, that science may be becoming more difficult.

Consider a related example from the computer industry. In 1965, Intel cofounder Gordon Moore noticed that the number of transistors per square inch on integrated circuit boards (now called microchips) had doubled every year since their invention. He predicted that this trend would continue into the foreseeable future. Moore's prediction proved correct: For nearly half a century, the chip density has indeed doubled approximately every 18 months, representing one of the most robust growth phenomena in history. But this growth required the efforts of an ever-increasing number of researchers [259]. Indeed, the number of individuals required to double chip density today is more than 18 times larger than the number required in the early 1970s. Therefore, even though processing power continues to grow exponentially, when it comes to producing the next generation of microchip, individual productivity has plummeted; every new release requires unprecedented amounts of manpower. Similar patterns have been observed in a wide array of industries, from agricultural crop yields to the number of new drugs approved per billion dollars spent on research [259]. This means that in these industries, the exponential growth of progress hides the fact that "apples" are indeed getting harder and harder to find.

In science, while the number of publications has grown exponentially, large-scale text analysis of physics, astronomy, and biomedicine publications revealed that the number of unique phrases in article titles has grown only linearly [260]. This suggests that the cognitive space of science – approximating the number of distinct scientific ideas in existence – may be growing much slower than scientific outputs.

So, is science running out of gas? As shown in this chapter, the exponential increase of scientific publications is indeed accompanied by an exponential growth in the number of scientists. But, if we compare the two growth rates across a large variety of disciplines, ranging from computer science to physics and chemistry to biomedicine, we find that the former is often comparable with the latter [4]. Indeed, as we discussed in Chapter 1, in science, individual productivity has stayed relatively stable over the past century, and even increased slightly in recent years.

As scientists look back upon the twentieth century, we marvel at the discoveries and inventions that our predecessors have made, from internal combustion engines to computers to antibiotics. The tremendous progress in these areas could imply diminishing returns, as the sweetest, juiciest apples are already harvested. Yet this does not seem to be the case. The data suggests that we are poised to discover and invent even more in the next 20 years than we have in all of scientific history. In other words, even after a century of breakneck progress, science today is fresher and more energetic than ever.

How can science continue to run tirelessly after a century of exponential growth? Because unlike a car or a colony of bacteria, science runs on ideas. While a car will eventually run out of gas, and bacteria will run out of nutrients, ideas are resources that grow the more they are used. Existing ideas give birth to new ones, which soon begin to multiply. So, while our ability to further improve internal combustion engines, computers, or antibiotics may indeed have diminished, we now look forward to new advances in genetic engineering, regenerative medicine, nanotechnology, and artificial intelligence – fields that will once again revolutionize science and our society, opening up whole new chapters beyond our wildest imagination.

16 CITATION DISPARITY

On the fiftieth anniversary of Eugene Garfield's *Science Citation Index (SCI)*, *Nature* teamed up with Thomson Reuters to tally the number of papers indexed by *SCI*, counting 58 million in all [261]. If we were to print just the first page of each of these papers and stack them on top of each other, the pile would almost reach to the top of Mt. Kilimanjaro (Fig. 16.1).

While this heap is certainly impressive, even more remarkable is the disparity in scientific impact that the mountain hides. If we order the pages based on the number of citations each paper received, placing the most cited ones on top and working downward, the bottom 2,500 meters – nearly half of the mountain – would consist of papers that have never been cited. At the other extreme, the top 1.5 meters would consist of papers that have received at least 1,000 citations. And just a centimeter and a half at the very tip of this mountain would have been cited more than 10,000 times, accounting for some of the most recognizable discoveries in the history of science.

16.1 Citation Distribution

The difference in impact among papers can be captured by a citation distribution, $P(c)$, representing the probability that a randomly chosen paper has c citations. de Solla Price was the first to compute this distribution [262], relying on citation data manually curated by Garfield and Sher in the early 1960s [263]. For this he counted by hand the number of papers that had been cited once, twice, three times, and so

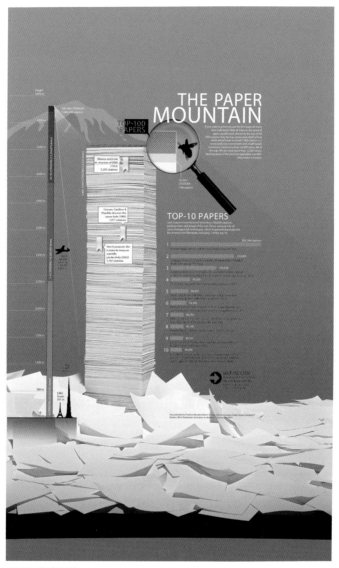

Figure 16.1 **The mountain of science.** If you print out just the first page of each research paper indexed in Web of Science by 2014, the stack would reach almost to the top of Mt. Kilimanjaro. Only the top 1.5 m of that stack would have received 1,000 citations or more, and just 1.5 cm would have been cited more than 10,000 times. The top 100 papers have all been cited more than 12,000 times, including some of the most recognizable scientific discoveries in history. After Van Noorden [261].

forth. He found that most papers had very few citations, but that a handful of papers had many more. Plotting the data, he realized that the citation distribution can be approximated by a power-law function,

$$P(c) \sim c^{-\gamma}, \tag{16.1}$$

with citation exponent $\gamma \approx 3$.

Most quantities in nature follow a normal or Gaussian distribution, commonly known as a bell curve. For example, if you measure the height of all your adult male acquaintances, you will find that most of them are between five and six feet tall. If you draw a bar graph to represent their heights, it will have a peak somewhere between five and six feet, and will decay quickly as you move away from the peak in either direction. After all, it is very unlikely that you know an eight-foot or a four-foot adult. Many phenomena, from the speed of molecules in a gas to human IQ, follow a similar bell curve.

Power laws like (16.1), however, belong to a different class of distributions, often called "fat-tailed distributions." The key property of these fat-tailed distributions is their high variance. For example, as Fig. 16.2 shows, the millions of papers with only a few citations coexist with a tiny minority of papers that have thousands of citations. (See also Box 16.1 The 80/20 rule.) If citations were to follow a bell curve, we would never observe such highly cited papers. Indeed, imagine a planet where the heights of the inhabitants follow a fat-tailed distribution. On

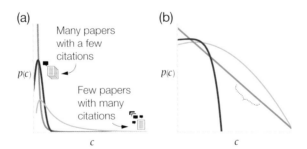

Figure 16.2 Illustrating normal, power-law, and lognormal distributions. (a) Comparing a power law and a lognormal function to a normal distribution on a linear–linear plot. (b) The same comparison shown on a log–log plot, helping us see the fundamental difference between the normal and fat-tailed distributions in the high citation regime. A power law follows a straight line on a log–log plot, its slope indicating the citation exponent γ. Sometimes, it can be difficult to tell lognormal and power laws apart, as they appear similar on a log–log plot.

such a planet, most creatures would be quite short, under one foot – but we would also occasionally encounter two-mile-tall monsters walking down the street. The strangeness of this imaginary planet highlights the stark difference between the fat-tail distribution followed by citations and the bell curves frequently seen in nature (Fig. 16.2).

> **Box 16.1 The 80/20 rule**
>
> Power law distributions are frequently seen in the context of income and wealth. Vilfredo Pareto, a nineteenth-century economist, noticed that in Italy a few wealthy individuals were earning most of the money, while the majority of the population earned rather small amounts. Looking more closely, he concluded that incomes follow a power law [264]. His finding is also known as the 80/20 rule: He observed that roughly 80 percent of money is earned by only 20 percent of the population.
>
> Versions of the 80/20 rule apply to many quantities that follow fat-tailed distributions [68, 265, 266]. For example, in business, 20 percent of sales reps typically generate 80 percent of total sales; on the World Wide Web, 80 percent of links lead to 15 percent of webpages; and in hospitals, 20 percent of patients account for 80 percent of healthcare spending.

16.2 Universality of Citation Distributions

There are remarkable differences in citations between disciplines. For example, biology papers regularly collect hundreds or even thousands of citations, while a highly influential math paper may struggle to collect a few dozen. These differences are illustrated in Fig. 16.3a, which depicts the distribution of citations for papers published in 1999 across several disciplines. As these plots show, the probability of an aerospace engineering paper gathering 100 citations is about 100 times smaller than the odds of a developmental biology paper reaching the same threshold. These systematic differences indicate that simply comparing the number of citations received by two papers in different disciplines would be meaningless. The less-cited aerospace engineering paper may have reported defining discoveries in that field, whereas a more-cited biology paper may have reported a merely incremental advance within its discipline.

To get a better idea of how influential a particular paper is, we can compare it to the average article in its discipline. Dividing a paper's

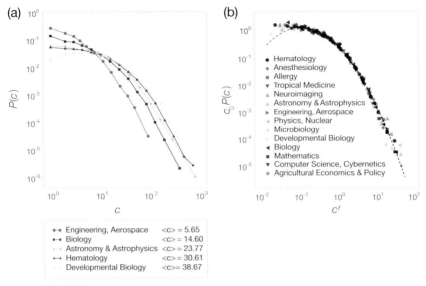

Figure 16.3 Universality of citation distribution. (a) The distribution of citations for papers published in 1999, grouped by discipline. The panel shows the probability $P(c)$ of a paper receiving exactly c citations for several scientific disciplines, illustrating that in some fields, like developmental biology, highly cited papers are more common than in engineering. (b) Rescaled citation distribution, illustrating that all curves shown in (a) follow the same distribution, once rescaled by $\langle c \rangle$, the average number of citations in the same field and year. The dashed line shows lognormal fit (16.2) to the data. After Radicchi et al. [43].

citations by the average number of citations for papers in the same field in the same year gives us a better measure of relative impact. When the raw citation counts are normalized in this way, we find that the distribution for every field now neatly follows a single universal function (Fig. 16.3b). The finding that the curves that were so visibly different in Fig. 16.3a now collapse in a single curve offers two important messages:

(1) Citation patterns are remarkably universal. Whether you publish in math, the social sciences or biology, the impact of your work *relative* to your own discipline has the same odds of being mediocre, average, or exceptional as the research happening in every other building on campus (Fig. 16.2b).

(2) The universal curves are well approximated by a lognormal function

$$P(c) \sim \frac{1}{\sqrt{2\pi}\sigma c} \exp\left(\frac{-(\ln c - \mu)^2}{2\sigma^2}\right). \tag{16.2}$$

> **Box 16.2 A variety of functions capture citation distributions**
>
> As illustrated in Fig. 16.2, when plotted, fat-tailed distributions like the lognormal (16.2) or power-law function (16.1) can appear quite similar to each other. But how do we know which function offers the best fit? Various studies have argued that a variety of functions could capture the citation distribution – from power laws [265–268] and shifted power laws, [269] to lognormals [43, 268, 270–275], to other more complicated forms [276–280]. How well these distributions suit the data often depends on which corpus of papers researchers analyzed, the publication year, the selection of journals, scholars, and their departments, universities, and nations of origin, to name a few factors. The mechanisms governing the emergence of different forms for $P(c)$ is an active research topic and will be discussed in the next chapters.

The universality of the citation distributions offers a simple way to compare scientific impact across disciplines. Let's focus on two hypothetical papers: Paper A is a computational geometry paper published in 1978, which has collected 32 citations to date; paper B is a biomedical paper published in 2002 with 100 citations. Although it may feel like an apples-to-oranges comparison, the relative citation function helps us compare the impact of the two papers. To do so, we first take all computational geometry papers published in 1978 and count their average citations. Similarly, we take all biomedical papers published in 2002, and calculate their corresponding average. Comparing the raw citation counts of paper A and B to the averages in their respective field and year provides a metric of *relative* impact for each, allowing us to make an unbiased comparison between the two.

In Table A2.1 (in Appendix A2), we calculate the average citation counts up to 2012 for all subject categories in 2004. As the table reveals, the more papers in a particular subject area, the more citations its average paper tends to receive – likely because there are more opportunities to be cited in that field. For instance, the biological sciences have some of the highest average citation counts, mostly in the 20s, whereas those in engineering and math have some of the lowest. There can be large differences even within the same subject. For example, biomaterials, a hot subfield of material science averages 23.02. Yet, the characterization and testing of materials, another subfield of material science, averages just 4.59. Interestingly, the number of papers in the two areas are not that different (2,082 vs. 1,239), suggesting that biomaterials papers are often cited in other disciplines.

Box 16.3 The widening citation gap

The 2008 Occupy Wall Street Movement, highlighted the fact that in the US, 1 percent of the population earns a disproportionate 15 percent of total income. This signals a profound income disparity, rooted in the fact that the income distribution follows a power-law distribution. Yet the debate about the 1 percent is less about the magnitude of its wealth than its trend over time: Income disparity has been skyrocketing for the last several decades (Fig. 16.4).

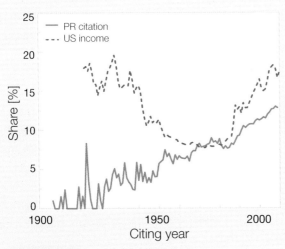

Figure 16.4 **The top 1 percent of Science.** The fraction of income earned by the top 1 percent of the population in US between 1930 and 2010, and the share of citations received by the top 1 percent most cited papers in *Physical Reviews (PR)*. We used the *PR* corpus of 463,348 papers published between 1893 and 2009 to examine the share of citations each paper gains the following year. After Barabási, Song, and Wang [281].

Given that citations follow fat-tailed distributions, could there be a similar rise in impact disparity in science? To answer this, we looked at the most-cited 1 percent of papers published in *Physical Reviews* journals in every year since 1893, measuring what fraction of total citations those papers received in the year following publication [281]. As Fig. 16.4 shows, impact disparity in the physical sciences has also been rising steadily over the past century.

16.3 What Do Citations (Not) Capture?

Citations and citation-based indicators are often used to quantify the impact or the quality of a paper. These metrics play an important role both in science policy and in the evaluation of individual scientists, institutions, and research fields. For example, citation counts are often discussed at hiring and tenure decisions, as well as in prize recommendations. Furthermore, many countries consider them in decisions pertaining to grants and other resources. Given their widespread use, it is easy to forget that citations are merely a *proxy* for impact or scientific quality. Indeed, there are many groundbreaking scientific discoveries that have received relatively few citations, even as less important papers amass hundreds. Consider some of the reasons why this could be:

- Review papers that summarize the state of a field or a topic, tend to be more frequently cited than regular articles, yet are often viewed as minor contributions to science than original research [282].
- Sometimes citations seek not to build upon what came before, but rather to criticize or correct previous papers. These should be viewed as "negative citations" – yet citation counts do not distinguish between supporting and critical references [283].
- Some citations may be "perfunctory," simply acknowledging that other studies have been conducted without contributing to the main narrative of the discovery. Indeed, manual examinations of 30 articles in *Physical Review* on theoretical high energy physics from 1968 to 1972 suggest that the fraction of such citations could be substantial [284]. This means, while each citation is counted the same, they have different roles in advancing science.

All of which prompts the question: Are better cited papers really better? To what degree do citations correctly approximate scientific impact, or what working scientists perceive as important? A number of studies have explored these questions. Overall, they find that citations correlate positively with other measures of scientific impact or recognition, including awards, reputation [285], peer ratings [286–290], as well as the authors' own assessments of their scientific contributions [282]. For example, a survey of the 400 most-cited biomedical scientists asked each of them a simple question [11]: Is your most highly cited paper also your most important one? The vast majority of this elite group answered yes, confirming that citation counts do constitute a valuable metric for a paper's perceived significance.

But would other researchers in the field agree? If we picked two papers in the same field, showed them side-by-side to researchers who work in that area, and asked them to say which is more relevant to their research, would they reliably pick the paper with more citations?

Close to 2,000 authors from two large universities performed precisely this exercise. However, the results were not straightforward [291]. The scholar's selection typically depended on whether or not he was asked to judge his own work. If both papers offered had been authored by the researcher, he systematically picked his better cited paper as the more relevant one. If, however, the author was asked to compare his paper with a paper by someone else, he overwhelmingly preferred his own paper – even if the other option was one of the most cited papers in their field, and even if the difference in impact amounts to several orders of magnitude.

These results suggest that, just as someone else's baby – no matter how adorable – can never compare to your own, scientists have a blind spot when asked to compare their work to that of others, which can skew their perception of even the most pivotal work.

Taken together, these surveys seem to tell two conflicting stories. On one end, scientists are not immune to biases, especially when it comes to their own work. But at the same time, citations could be a meaningful measure, since they largely align with the perceptions of experts. This suggests that, despite their shortcomings, citations likely play a critical role in gauging scientific impact.

But there is another, more fundamental reason why citations matter: There is not one scientist in the world who can single-handedly demand that a paper amass citations [16]. Each of us decides on our own whose shoulders to stand upon, citing the papers that inspired our work, and those which were key to developing our ideas. When all of those individual decisions are combined, what emerges is the collective wisdom of the scientific community on a paper's importance. It is not enough for one scientist to decide that a paper is great. Others in the community must also agree, with each additional citation acting as an independent endorsement. In other words, scientific impact is not about what *you* think; it's about what *everyone else* thinks.

Box 16.4 Patent citations

Citations can also help quantify the importance of inventions. When a patent is issued by the patent office, it includes citations to prior patents, as well as to the academic literature. Patent citations can be added by inventors, their attorneys or by patent examiners. Just like in science, patents with the most citations tend to describe important inventions [292, 293]. But patent citations go a step further, conveying not just an invention's importance, but also its economic impact. For example, studies have found that more highly cited medical diagnostic imaging patents produce machines that attract more demand [294] and that companies with highly cited patents have higher stock market values [295]. More importantly, the link between the commercial value of a patent and its citations is positive and highly nonlinear. One survey polled patent owners 20 years after their invention, asking them how much they should have asked for their patent knowing what they know today. The responses revealed a dramatic, exponential relationship between the economic value of a patent and its citations [296]: A patent with 14 citations, for example, had 100 times the value of a patent with 8 citations.

17 HIGH-IMPACT PAPERS

Why are most papers rarely cited, and how do a few lucky ones turn into runaway successes? These questions present a deep puzzle. In every new paper, the authors carefully select which work to cite, based on the topic of the manuscript and their own familiarity with the literature. Yet somehow these many personal decisions result in a highly stable citation distribution, capturing impact disparities that transcend disciplines. How do citation superstars emerge? What determines which papers are highly cited, and which are forgotten? And why are these universal citation distributions universal, independent of the discipline?

In this chapter we show that, perhaps somewhat counterintuitively, citation superstars and the universality of the citation distribution emerge precisely *because* citations are driven by individual preferences and decisions. Although our individual choices differ widely, the behavior of the scientific community as a whole follows highly reproducible patterns. As such, despite the seemingly countless factors that govern each paper's impact, the emergence of exceptionally impactful papers can be explained by a handful of simple mechanisms.

17.1 The 'Rich-Get-Richer' Phenomenon

No scientist can read the million or so scientific papers published each year. So we typically discover papers of interest while reading other papers, and looking at the work they cite. This leads to a curious bias: the more widely cited a paper already happens to be, the more likely it is that we will encounter it through our reading. And since we typically only cite work that we read, our reference list is frequently populated with highly cited papers. This is an example of a "rich-get-richer" phenomenon – similar to the Matthew effect encountered in Chapter 3 – the higher a paper's citation count, the more likely it is to be cited again in the future.

As simple as it seems, the rich-get-richer mechanism alone can explain much of the citation disparity among scientific publications, and for the universal, field-independent nature of citation distributions. This was formalized in a model first proposed by de Solla Price in 1976 (Box 17.1), sometimes called the *Price model* [297, 298], which incorporates two key aspects of citations:

(1) **The growth of the scientific literature.** New papers are continuously published, each of which cite a certain number of previous papers.
(2) **Preferential attachment.** The probability that an author chooses a particular paper to cite is not uniform, but proportional to how many citations the paper already has.

As discussed in more detail in Appendix A2.1, the model with these two ingredients (growth and preferential attachment) predicts that citations follow a power-law distribution, and hence can explain the empirically observed fat-tailed nature of citations.

Box 17.1 The rich-get-richer effect

The rich-get-richer effect has been independently discovered in multiple disciplines over the last century, helping to explain disparities in city and firm sizes, abundance of species, income, word frequencies, and more [60, 297, 299–304]. The best-known version was introduced in the context of complex networks [67, 68], where the term preferential attachment was proposed within the context of the Barabási–Albert model [304] to explain the existence of hubs in real networks. In sociology, it is often called "the Matthew effect," as discussed in Chapter 3 and also called "cumulative advantage" by de Solla Price [297].

This analysis leads us to a critical takeaway. Clearly there are myriad factors that contribute to the impact of a paper, some of which will be considered in later chapters. But the accuracy with which Price's model captures the empirically observed citation distribution demonstrates that these additional factors don't matter if our only goal is to explain the origin of the fat-tailed distribution of citations. Growth and preferential attachment, together leading to a rich-get-richer effect, can fully account for the observed citation disparity, pinpointing the reason why citation superstars emerge (see Appendix A2.2 for the origins of preferential attachment). It also explains why citation distributions are so universal across widely different disciplines: While many factors may differ from one discipline to the other, as long as the same preferential attachment is present, it will generate a similar citation distribution, independent of any disciplinary peculiarities.

17.2 First-Mover Advantage

Price's model has another important takeaway: The older a paper is, the more citations it should acquire. This phenomenon is called the *first-mover advantage* in the business literature. That is, the first papers to appear in a field have a better chance of accumulating citations than papers published later. Then, thanks to preferential attachment, these early papers will retain their advantage in perpetuity.

This prediction was put to the test by analyzing citations patterns of papers published in several subfields, like network science and adult neural stem cells [305], finding a clear first-mover effect, whose magnitude and duration are quite similar to that predicted by Price's model. To appreciate the size of this effect, let's look at the field of network science, born at the end of the 1990s. The first 10 percent of papers published in this field received on average 101 citations, while the second 10 percent received just 26 on average. Since the second batch was published immediately after the first, the difference is likely not due to the fact that the second batch had less time to collect citations.

Scientists tend to treat the first papers as founders of a field, which may explain their high citation count. The first researcher to bring a problem to the attention of the scientific community deserves credit for it, regardless of whether all of the original results retain their relevance.

Occasionally, however, prominent latecomers do take over. Consider, for example, the Bardeen–Cooper–Schrieffer (BCS) paper [306], which introduced the first widely accepted theory of superconductivity. It was a relatively late contribution to the field of superconductivity. Yet by explaining a wide range of puzzles plaguing the field with a single elegant theory, it quickly became the defining paper, and the citations followed. Hence, the BCS paper is a living testament to the fact that the first-mover principle is not absolute, prompting us to ask: If only the rich are allowed to get richer, how can a latecomer succeed?

To answer this question, we must recognize that preferential attachment is not the only mechanism driving citation counts. As we will see next, there are other mechanisms that determine which papers loom large in the scientific canon, and which fade into obscurity.

17.3 The Fit Get Richer

Despite its simplicity, Price's model omits a crucial factor that does influence the accumulation of citations: Not all papers make equally important contributions to the literature. Indeed, there are huge differences in the perceived novelty, importance, and quality of papers. For example, some papers report highly surprising discoveries that alter the prevailing paradigm, and others, like the BCS paper, bring clarity to long-standing puzzles in a large and active field – but these papers coexist with other publications that merely rehash old ideas or proffer half-baked theories. Papers also differ in their publication venues, the size of the audience they speak to, and the nature of their contribution (i.e., review papers and method papers tend to be cited more than regular research papers). In other words, papers differ in their inherent ability to acquire citations, each being characterized by some set of intrinsic properties that will determine its impact relative to the pack. We will call this set of properties *fitness*, a concept borrowed from ecology and network science [307].

Price's model assumes that the growth rate of a paper's citations is determined solely by its current number of citations. To build upon this basic model, let's assume that citation rate is driven by both preferential attachment and a paper's fitness. This is called the *fitness model* [307, 308], or the Bianconi–Barabási model, which incorporates the following two ingredients. (See Appendix A 2.3 for more detail.)

(1) **Growth**: In each time step, a new paper i with a certain number of references and fitness η_i is published, where η_i is a random number chosen from a distribution $p(\eta)$. Once assigned, the paper's fitness does not change over time.
(2) **Preferential attachment**: The probability that the new paper cites an existing paper is proportional to the product of paper i's previous citations and its fitness η_i.

Here a paper's citation is not just dependent on its existing citations, captured by the preferential attachment mechanism we have discussed earlier. It also depends on its fitness, indicating that between two papers with the same number of citations, the one with higher fitness will attract citations at a higher rate. Hence, the presence of fitness assures that even a relatively new paper, with a few citations initially, can acquire citations rapidly if it has greater fitness than other papers.

> **Box 17.2 The origin of the lognormal citation distribution**
>
> As we discussed in Chapter 16, several recent studies have indicated that citation distributions sometimes are better fitted by lognormal distributions. While a lognormal form is inconsistent with the Price's model (which predicts a power law), it can be explained by the fitness model [307]. Indeed, if every paper has the same fitness, thanks to preferential attachment, citations will follow a power law distribution. If, however, papers differ in their fitness, then the underlying distribution of fitness parameter $p(\eta)$, from which we draw a number to assign a paper its fitness, determines the shape of citation distribution. For instance, if $p(\eta)$ follows a normal distribution, which is a natural assumption for bounded quantities like fitness, the fitness model predicts that the citation distribution should follow a lognormal function [273]. As we show in Chapter 18, where we estimate the fitness of a paper, the fitness distribution is indeed bounded, like a normal distribution. This suggests that if we wish to explain the empirically observed citation distributions, we cannot ignore the fact that papers have different fitness.

Differences in fitness explain how a latecomer can overtake established citation leaders. Indeed, a strong paper may appear late in the game but nevertheless grab an extraordinary number of citations within a short time frame. Unlike Price's model (which predicted that citations of all papers grow at the same rate), the fitness model predicts that the growth rate of citations is proportional to the paper's fitness, η. Thus, a paper with higher fitness acquires citations at a higher rate, and given sufficient time, it will leave behind the older papers with lower fitness.

But what does it take to write a high fitness paper?

18 SCIENTIFIC IMPACT

If we assume that citations approximate a paper's scientific impact, then the fat-tailed shape of the impact distribution implies that most papers have, unfortunately, almost no impact at all; indeed, only a very small fraction of the literature affects the development of a field. As the preceding chapter showed, high fitness is crucial for certain ideas to make a large impact. But what predicts high fitness? And how can we amplify the scientific impact of our work? In this chapter, we will focus on the role of two different factors – one is internal to a paper and the other is external – novelty and publicity.

18.1 The Link between Novelty and Scientific Impact

While many qualities can affect the 'fitness' of a paper, one, in particular, has attracted much attention: novelty. What exactly is novelty, and how do we measure it in science? And does novelty help or hurt a paper's impact?

18.1.1 Measuring Novelty

As discussed in Chapter 3.1, new ideas typically synthesize existing knowledge. For example, inventions bring together pre-existing ideas or processes to create something original (Fig. 18.1) [309]. The steamboat is a combination of a sailing ship and a steam engine, and the Benz Patent-Motorwagen, the first automobile in the world, combined a bicycle, a carriage, and an internal combustion engine. Even the

Figure 18.1 New ideas are often an original combination of existing ones. The Benz Patent-Motorwagen ("patent motorcar"), built in 1885, is regarded as the world's first production automobile. The vehicle was awarded the German patent number 37435, for which Karl Benz applied on January 29, 1886. The Motorwagen represents a combination of three pre-existing ideas: bicycle, carriage, and internal combustion engine.

smartphone in your pocket is simply a combination of many pre-existing parts and features: memory, digital music, a cell phone, Internet access, and a lightweight battery.

The theory that existing technologies are recombined to generate new inventions is confirmed by the analysis of US patents [310]. Each patent is classified by the US patent office (USPTO) using a unified scheme of technology codes (a class and a subclass). For example, one of the original patents for iPod, assigned to Apple Computer, Inc with Steve Jobs listed as one of the inventors [US20030095096A1], has a class-subclass pair 345/156, denoting class 345 (Computer Graphics Processing and Selective Visual Display Systems) and subclass 156 (Display Peripheral Interface Input Device). Examining all US patents dating from 1790 to 2010, researchers found that, during the nineteenth century, nearly half of all patents issued in the US were for single-code inventions – those that utilize a single technology, rather than combining multiple technology areas. Today, by contrast, 90 percent of inventions combine at least two codes, showing that invention is increasingly a combinatorial process.

This combinatorial view of innovation offers a way to quantify novelty in science. Indeed, scientific papers draw their references from multiple journals, signaling the domains from which they sourced their ideas [92, 311, 312]. Some of these combinations are anticipated, whereas others are novel, deviating from conventional wisdom.

If a paper cites a pair of journals that are rarely brought together, it may suggest that the paper introduces a novel combination of prior work. Take for instance a 2001 paper in the *Journal of Biological Chemistry*, which pinpointed the protein with which a known antipsychotic drug interacts and used this insight to identify other biological effects [313]. Its reference list is the first ever to cite both the journal *Gene Expression* and the *Journal of Clinical Psychiatry* [314], representing a novel combination of prior knowledge. On the other hand, other journals cited in the same paper, like the *Journal of Biological Chemistry* and *Biochemical Journal*, are frequently cocited in the literature, an example of the conventional pairings that reflect more mainstream thinking in the field.

18.1.2 The Novelty Paradox

Evidence from a broad array of investigations consistently shows that rare combinations in scientific publications or patented inventions are associated with a higher likelihood that the publication or invention will achieve high impact. In other words, with novelty comes an increased chance of hitting a home run. This finding also validates the key premise of interdisciplinary research [315–317] – that many fruitful discoveries come from the cross-pollination of different fields and ways of thinking, combining previously disconnected ideas and resources [5, 317, 318].

Yet, while novel ideas often to lead to high-impact work, they also lead to higher degrees of uncertainty [311, 319, 320]. In fact, very novel ideas and combinations can just as well lead to failure as to a breakthrough. For example, an analysis of more than 17,000 patents found that the greater the divergence between the collaborators' fields of expertise, the higher the variance of the outcomes; highly divergent patents were more likely than average both to be a breakthrough, and to be a failure (Fig. 18.2) [320].

Similarly, papers that cite more novel combinations of journals are more likely to be in the top 1 percent of cited papers in their field. Yet at the same time, they are also riskier, tending to take a longer time before they begin accumulating more citations [311]. The higher risk

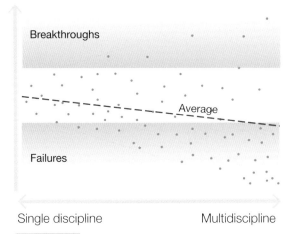

Figure 18.2 **Multidisciplinary collaborations in patenting.** As collaborations among inventors become more multidisciplinary, the overall quality of their patents decreases. But multidisciplinary collaboration increases the variance of the outcome, meaning that both failures and breakthroughs are more likely. After Fleming [320].

inherent in innovation may play a major role in determining what kind of innovation takes place (or doesn't) in academia. For instance, in the field of biochemistry, studying chemical relationships between unexplored compound pairs is much more novel than focusing on well-studied chemicals, and such strategies are indeed more likely to achieve high impact. But the risk of failure of exploring such previously unexplored combinations is so high that, as an analysis estimated, the additional reward may not justify the risk [319].

The high variance in the impact of novel ideas may be rooted in the human bias against novelty. Studies of grant applications show that scientists tend to be biased against novel concepts before the work is realized. At a leading US medical school, researchers randomly assigned 142 world-class scientists to review 15 grant proposals. In parallel, the researchers measured the occurrences of rare combinations of keywords in each proposal [321]. For example, proposals combining the terms "Type 1 diabetes" and "Insulin" were typical, whereas proposals with "Type 1 diabetes" and "Zebrafish" presented a novel combination rarely seen in the literature. But would the more novel proposals be graded more or less favorably? The researchers found that proposals that scored high on novelty received systematically lower ratings than their less novel counterparts. Even nominally "interdisciplinary" grants are not immune to similar biases [322]. Analyzing all 18,476 proposals submitted to an Australian funding program, including both successful and unsuccessful applications, researchers measured how many different fields were represented in each proposal, which is weighted by how distant those fields were. The results indicated that the more interdisciplinary the proposed work, the lower the likelihood of being funded.

And, so we are left with a paradox. It is clear that novelty is essential in science – novel ideas are those that score big. Yet the novelty bias observed in grant applications suggests that an innovative scientist may have trouble getting the funding necessary to test these ideas at the first place. And, even if she does, novel ideas are more likely to fail than mediocre ones.

Is there anything we can do to ameliorate this paradox? Recent studies have offered one crucial insight: balance novelty with conventionality. Consider that Darwin devoted the first part of *On the Origin of Species* to highly conventional, well-accepted knowledge about the selective breeding of dogs, cattle, and birds. In doing so, he exhibited an essential feature of many high-fitness ideas that do achieve great impact: They tend to be grounded in conventional combinations of prior work, while also merging hitherto un-combined, atypical knowledge. Analyzing 17.9 million papers spanning all scientific fields, researchers found that papers that introduced novel combinations, yet remained embedded in conventional work, were at least twice as likely to be hits than the average paper [92]. These results show that novelty can become especially influential when paired with familiar, conventional thought [92, 323].

18.2 Publicity (Good or Bad) Amplifies Citations

Does media coverage amplify scientific impact? Are we more likely to cite papers that have been publicized in the popular press? To answer these questions, let's turn to a major news outlet: *The New York Times*.

Since papers about human health are often of general interest, researchers in one study looked at whether articles published by *The New England Journal of Medicine* (*NEJM*) were covered by the *Times*. They compared the number of citations *NEJM* articles received when they had been written up in the *Times*, versus when they were not written up in the *Times*. [324]. They found that overall, articles covered by the *Times* received 72.8 percent more citations in the first year than the non-covered group.

But can we attribute this dramatic impact difference to the publicity that the *Times* offers? Or could it be that the *Times* simply covered outstanding papers, which would have gathered just as many citations without their coverage? A natural experiment allowed the researchers to find a more definitive answer: The *Times* staff underwent a 12-week strike from August 10 to November 5, 1978. During this period, it continued to print a reduced "edition of record" but did not sell copies to the public. In other words, it continued to earmark articles it deemed worthy of coverage during the strike, but this information never reached its readership. During this period, researchers found, the citation advantage disappeared entirely – the articles that the *Times* selected for coverage did no better than those it didn't in terms of citations. Therefore, the citation advantage of attention-grabbing papers cannot be explained solely by their higher quality, novelty, or even mass appeal – it is also the result of media coverage itself.

It is not hard to see why publicity helps boost citations. Media coverage increases the reach of the audience, potentially allowing a wider group of researchers to learn about the findings. It may also act as a stamp of approval, bolstering the credibility of the paper in the eye of the scientific community. But perhaps the most basic reason is that media publicity is, more often than not, *good* publicity. Indeed, a TV station or newspaper does not pretend to be a check or balance on science. When media chooses to spend its limited air time or ink on a scientific study, it usually presents findings that are deemed genuine, interesting, and important – after all, if they weren't all of these things, then why bother wasting the audience's time?

Media offers only good publicity for science, which may have important consequences for the public perception of science (see Box 18.1). But, the checks and balances used to ensure that scientific work is accurate and honest are maintained by scientists. Scientific critiques and rebuttals can come in many forms: some only offer an alternative interpretation of the original results, and others may refute only a part of a study. In most cases, however, rebuttals aim to highlight substantial flaws in published papers, acting as the first line of defense after scientific research has passed through the peer review

Box 18.1 Media bias and science

The important role the media plays in disseminating science raises a critical question: Does the press offer balanced coverage? Research on media coverage of medical studies found that journalists prefer to cover only initial findings, many of which are refuted by subsequent studies and meta-analyses. But journalists often fail to inform the public when the studies they covered are disproven [325, 326]. When a team of researchers examined 5,029 articles about risk factors for diseases and how those articles were covered in the media [326], they found that studies reporting positive associations about disease risk and protection (e.g., studies suggesting that a certain food may cause cancer, or that a certain behavior may help stave off heart disease) tend to be widely covered. In contrast, studies finding no significant association received basically zero media interest. Moreover, when follow-up analyses fail to replicate widely reported positive associations, these follow-up results are rarely mentioned in the media. This is troubling since the researchers found that, of the 156 studies reported by newspapers that initially described positive associations, only 48.7 percent were supported by subsequent studies.

This hints at the main tension between the media and the sciences: While the media tends to cover the latest advances, in science, it's the complete body of scientific work that matters. A single study can almost never definitively prove or disprove an effect, nor confirm or discredit an explanation [327, 328]. Indeed, the more novel a paper's initial findings, the more vulnerable they are to refutation [8].

The media's tendency to report simple, preliminary results can have severe consequences, as the media coverage of vaccinations reveals. Presently, many parents in the United States refuse to vaccinate their children against measles, mumps, and rubella (MMR), fearing that the vaccine could cause autism. Why do they believe that? This fear is rooted in a 1998 paper published by Andrew Wakefield in *The Lancet*, which received worldwide media coverage [329]. However, the original study was based on only 12 children, and follow-up studies unanimously failed to confirm the link. Furthermore, researchers later found that Wakefield had distorted his data, hence the paper was retracted. He subsequently lost his license and was barred from practicing medicine (see https://en.wikipedia.org/wiki/Andrew_Wakefield). Yet, although Wakefield's findings have been widely refuted and discredited in the scientific community, the media's coverage of Wakefield's claim led to a decline in vaccination rates in the United States, the United Kingdom, and Ireland.

system. Here, it seems, we finally have a form of *bad* publicity. But do these critiques and rebuttals diminish a paper's impact? And, if so, how much?

Comments, which tend to question the validity of a paper, are often seen as "negative citations," ostensibly making the original paper less trustworthy in the eye of the scientific community. Hence, one would expect commented papers to have less impact. Yet studies have revealed the opposite: Commented papers are not only cited more than non-commented papers – they are also significantly more likely to be among the most cited papers in a journal [283].

Similar results are uncovered by studies of negative citations – references that pinpoint limitations, inconsistencies, or flaws of prior studies [330]. Researchers used machine learning and natural language processing techniques to classify negative citations from a training set of 15,000 citations extracted from the *Journal of Immunology*, categorizing the citations as "negative" or "objective" with the help of five immunology experts. They then used the tool to analyze 15,731 articles from the same journal. They found that papers pay only a slight long-term penalty in the total number of citations they receive after a negative one, and the criticized papers continue to garner citations over time, which paints the picture that it's better to receive negative attention than none at all.

Together, these results show that comments and negative citations seem to play a role that's the opposite of what is intended; they are early indicators of a paper's impact. Why does such bad publicity amplify citation impact?

The main culprit is a selection effect: Scientists are often reluctant to devote time to writing comments on weak or irrelevant results [331]. Hence, only papers perceived as potentially significant draw enough attention to be commented on in the first place. Moreover, while comments or negative citations are critical in tone, they often offer a more nuanced understanding of the results, advancing the argument presented in the paper rather than simply invalidating its key findings. In addition, comments also bring attention to the paper, further boosting its visibility. Even in science, it appears, there's no such thing as bad publicity.

19 THE TIME DIMENSION OF SCIENCE

The Library of Alexandria had a simple but ambitious goal: to collect all knowledge in existence at the time. Built in the Eastern Harbor of Alexandria in Egypt, where Queen Cleopatra first laid eyes on Julius Caesar, the library relied on a unique method to enrich its collection: Every ship that sailed into the busy Alexandria harbor was searched for books. If one was found, it was confiscated on the spot, and taken to the library where scribes would copy it word for word. The rightful owner was eventually returned the copy along with adequate compensation, while the library kept the original. Although historians continue to debate the precise number of books it managed to amass, the library at its peak was estimated to possess nearly half a million scrolls – that is, until Julius Caesar burned it all down in 48 BCE.

Imagine you are at the library of Alexandria, watching the fire spread slowly towards the building. You are surrounded by the world's single greatest archive of knowledge and you know that it will soon be turned into ashes. You have time to step in and save a few scrolls. Which ones should you rescue? Should you collect the oldest scrolls, containing the ideas that have stood the test of time? Or should you run for the most recent documents, as they synthesize the best of past knowledge? Or perhaps you should pick some documents at random. The fire is approaching, and time is running out – what would you do?

Tomorrow's biggest discoveries necessarily build on past knowledge. But when choosing which discoveries to build upon, how far back should we go? And, relatedly, how long will others continue to cite our own work before the fires of time render it irrelevant? These are the questions we will explore in this chapter. We will pinpoint the unique combinations of old and relatively new knowledge that are most likely to produce new breakthroughs. In doing so, we will see that the way we build on past knowledge follows clear patterns – and we will explore how these patterns shape future scientific discourse. Let's begin with a simple question: How far back into the past do scientists typically anchor their inquiries?

19.1 Myopic or Not?

The famous Newton quote, "If I have seen further than others, it is by standing upon the shoulders of giants," suggests that older work, tested by time, is the foundation for new discoveries. However, Francis Bacon disagreed, arguing that discoveries occur only when their time has come [245]. If Bacon is right, then it is the most *recent* work that drives breakthroughs. So, if you don't want to miss out on a discovery, you'd better stay at the bleeding edge of knowledge.

Who is right, Newton or Bacon? One way to test the two theories is to compile the age distribution of the references in research papers, measuring the time gap between a work's publication year and the publication years of its references. This approach has a long history: librarians used to look at the ages of references to determine which older journal volumes could be discarded to free up shelf space [332].

Figure 19.1 shows the probability that a paper cites a paper published t years earlier. The distribution vividly documents the myopic nature of science: most references point to works published just two to three years earlier. One may argue that this is a consequence of the fact that there were many more papers published in recent years than in years further back [297, 333]. But even when we account for the exponential growth of literature, the distribution nonetheless decays rapidly with time. While scholars do regularly reach back to "vintage" knowledge more than 20 years old, the likelihood of citing research older than that decays rather quickly.

The distribution in Fig. 19.1 captures the collective foraging patterns of scientists, showing that we tend to rely heavily on recent information, yet balance it occasionally with vintage, or canonical knowledge. Individual scientists may differ, however, in how they search for past knowledge, raising the question: Could the varied

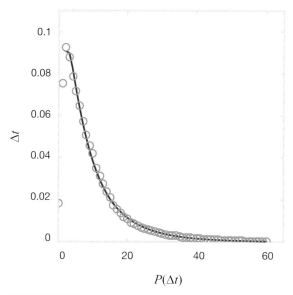

Figure 19.1 **The age of citations.** To understand where new knowledge comes from, we can look at how old the references of a paper are. The age distribution of citations for papers published in 2010 shows a peak around two years and decays quickly afterwards. Much research has been devoted to reasoning about how best to characterize the distribution of citation ages. A classic meta-analysis of over 20 studies indicates that the age of references is best approximated by an exponential distribution [297]. A more recent analysis suggests that a lognormal distribution with an exponential cutoff offers a better fit [253], shown in the figure as a solid line. After Yin and Wang [253].

patterns of information foraging predict the impact of their work? Is there a search strategy that is particularly well-suited to prompting tomorrow's breakthroughs?

19.2 The Hotspot of Discovery

Imagine four papers that differ in the way they cite previous literature (Fig. 19.2). Which one will have the highest impact? We can make a case for each of them. The deep-reach paper (Fig. 19.2a) primarily cites old papers, building on well-established, well-tested classics. By contrast, the paper in Fig. 19.2b draws only upon hot, recent topics that are absorbing the community's attention at the moment. The paper in Fig. 19.2c combines the best of both worlds, featuring a judicious mix of new and vintage knowledge, with an emphasis on more recent

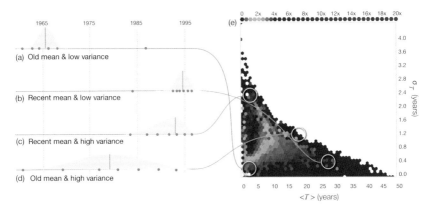

Figure 19.2 Knowledge hotspots predict high impact. (a–d) Potential information search patterns characterizing individual papers. (e) The hotspot, shown in red, captures "hit" papers that cite prior knowledge with a low mean age, $<T>$, and a high coefficient of variation for age, σ_T (data shown for the year 1995, capturing $N = 546,912$ publications). The background rate is the likelihood that a paper chosen at random is in the top 5 percent of citations for papers in its field. Papers in the hotspot are on average more than two times more likely than the background rate to be hits. Notably, 75 percent of papers are outside the hotspot, and their likelihood of being a hit is no greater than expected by chance. After Mukherjee et al. [334].

discoveries, while the one in Fig. 19.2d draws upon knowledge somewhat uniformly across time.

Which strategy is associated with the greatest impact? To answer this question, a team of researchers analyzed more than 28 million papers in the Web of Science, capturing two parameters for each paper [334]: the mean age of its references ($<T>$) and the coefficient of variation (σ_T) for the age distribution of its references, which captures how far those references' publication dates are spread out. Then the researchers measured the number of citations each paper had accumulated eight years after its publication. They considered a work of high-impact (i.e., a hit paper) if it lay within the top fifth percentile of citation counts in its subfield.

The relationship between $<T>$, σ_T, and impact is captured by the heat plot shown in Fig. 19.2e. Each point represents the $<T>$ and σ_T values of papers published in 1995 and the color corresponds to the probability of finding hit papers at that particular combination of T_μ and σ_T, defining the "hotspots" of discovery. Three main conclusions emerge from Fig. 19.2e:

- Papers with low $<T>$ and high σ_T are in the middle of the hotspot, offering the highest impact. In other words, papers whose references are centered just behind the bleeding edge of knowledge but which also combine an unusually broad mix of old and new (as in Fig. 19.2c) tend to garner the most citations. Papers of this sort are 2.2 times more likely to become homeruns in their field than a randomly chosen publication.
- Papers that center their references exclusively on new knowledge (those with low $<T>$ and low σ_T, Fig. 19.2b) have a surprisingly low impact, rarely exceeding what would be expected by chance. This suggests, the conventional bias towards heavily referencing recent work may be misguided; more recent references are only valuable when supplemented with older references.
- Papers whose references are centered in "vintage" knowledge (papers with high $<T>$ and low σ_T, Fig. 19.2a) have a particularly low impact. Their likelihood of becoming homeruns in their field is about half of what is expected by chance. This is quite remarkable, given that some 27 percent of all papers fall in this category.

Figure 19.2e shows that being at the cutting edge is key for future impact, but only when paired with an appreciation for older work. Building only on recent knowledge amounts to throwing a dart to determine your paper's impact – you are leaving it entirely to the whims of chance. Furthermore, the lowest impact works are those that are stuck in the past, building exclusively on vintage knowledge while oblivious of recent developments. These are not merely sweeping generalizations: Disaggregating the data by field reveals the same pattern of hotspots in nearly every branch of science. What's more, the effect of referencing patterns has become more universal in recent decades. While at the beginning of the postwar era about 60 percent of fields displayed a clear link between hotspots and hits, by the 2000s, this link was detectable in almost 90 percent of research fields.

This suggests, despite important differences among scientific fields in methods, culture, and the use of data versus theory, the way a scientist chooses to build upon earlier work consistently influences whether or not her work will make a splash, no matter her field.

19.3 The Growing Impact of Older Discoveries

Access to scholarly knowledge has fundamentally changed over the years. Today's students may never set foot in a library. Instead, they hunt for the relevant literature from their browsers, using powerful search engines like Google Scholar. And while they may be impressed by the heavy bookshelves and neat piles of printed journals in a professor's office, it may be hard for them to imagine how he once relied on them to keep up with the latest developments in his field. Many scientists today first learn about their colleagues' new research from social media and they feel satisfied when someone retweets their latest preprints.

But how do the changes in information access alter the way we build on past knowledge? On one hand, new tools and technologies allow us to reach deeper into the past [335–337]. Today every journal offers online access and a digital archive of its older articles, making all research accessible 24/7 for anyone with Internet access. These changes make older knowledge more accessible [337], ostensibly increasing the chances that scientists will build upon it, especially since search engines often allow users to see the most relevant results first, not just the most recent ones.

Yet there are also reasons to believe that these changes may in fact narrow the ideas scientists build upon [338]. Following hyperlinks is likely to steer researchers towards the current prevailing opinion, rather than something deep within the archives. Indeed, trips to the library might have been inefficient, but they also offered an opportunity for serendipity; flipping through dusty pages, scholars were forced to interact with findings from the distant past, noticing papers and knowledge they were not necessarily seeking. Therefore, digital publishing has the potential to make us more myopic, its convenience shifting us away from the classics and towards more recent work. Moreover, the digital tools we rely on also push us closer than ever to the frontier of knowledge. Today's scientist receives an email alert the moment a new paper in her field has been published or is shared on social media. She does not even need to wait for a journal publication to learn about the latest advances, as preprint sharing (wherein a paper is circulated publicly before being published in a peer-reviewed journal) is now the norm in several disciplines. Taken together, these changes may further reduce the time gap between what scientists produce and what they build upon.

To estimate the impact of digitization, researchers studied papers published between 1990 and 2013, counting how many cited papers that were at least 10 years old [337]. They found that the fraction of older citations grew steadily between 1990 and 2013, even accelerating after 2000. The shift was most pronounced between 2002 and 2013, when sophisticated search engines emerged. By 2013, four of the nine broad research areas had at least 40 percent citations to older articles, the "humanities, literature and arts" category leading the pack with 51 percent.

This trend is not limited to the last two decades. Plotting the average age of references over time, we find that since the 1960s, scientists have been systematically reaching deeper into the literature, citing older and older papers (Fig. 19.3). Why did such deep referencing start in the 1960s? While the answer is not entirely clear, one possible answer is rooted in the advent of peer review [5]. Before the 1960s,

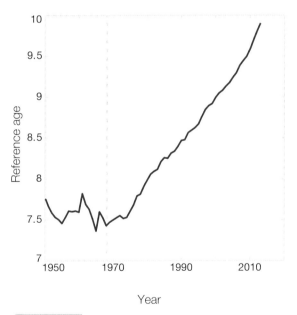

Figure 19.3 **Building on the past.** The average age of the references of papers stayed around 7.5 years from 1950 to 1970, after which we observe a remarkable increasing trend, indicating that scientists are systematically reaching deeper into the literature. The dotted line marks the year 1967, when *Nature* officially introduced peer review. The fact that the start of peer review coincided with changes in referencing patterns has led researchers to hypothesize a link between the two [5].

papers were accepted mainly at the editors' discretion [339]. By the mid-twentieth century the increasing specialization of science had intensified the need for expert opinions – yet disseminating a paper to geographically dispersed readers was difficult until photocopying became available in 1959 [340]. *Nature* officially introduced peer review in 1967, and the practice became mainstream throughout scholarly publishing shortly thereafter. It is likely that these reviewers could point out relevant older work for the authors to consider, not only boosting references to canonical literature, but also inducing a behavioral change, prompting authors to more carefully credit earlier work.

Together, these results indicate that technological shifts have made it easier to find the most relevant articles, irrespective of their age. Happily, important advances are no longer getting lost on shelves, but are able to continue to influence research for decades longer. But just how long can a researcher expect their corpus of work to remain relevant?

> **Box 19.1 The age distribution of citations: Two distinct approaches**
>
> Historically, two rather distinct approaches have been used to measure the age distribution of citations. The *retrospective (citation from)* approach considers papers cited by a publication during a particular year and analyzes the age distribution of these citations [272], looking back in time [341–343]. By contrast, the *prospective (citations to)* approach studies the distribution of citations gained over time by papers published in a given year [272, 341–344].
>
> Research shows that the two approaches are connected through precise mathematical relationships, allowing us to derive and even predict one approach from the other [253]. Yet, while both approaches measure the age of citations, they capture different processes. The retrospective approach, which we have used so far in this chapter, measures how far back in time the authors of a paper look as they cite earlier references, characterizing the citing "memory" of a paper. The prospective approach, on the other hand, represents a collective measure, quantifying how a particular paper is remembered over time. Hence, the prospective approach (which we will use next) is more useful for understanding how impact changes with time. Note, however, that studies prior to 2000 mostly relied on the retrospective approach. The reason is mainly computational: the retrospective approach is easier to implement, since the *citing* paper is fixed.

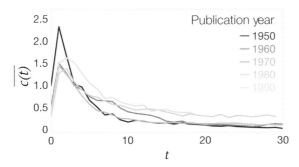

Figure 19.4 **The jump-decay pattern of citations.** The average number of citations as a function of the time elapsed since publication. Each line represents a group of papers, the color denoting their publication year. As the curves indicate, most citations come within the first five years after publication; after that, the chance of a paper being cited again drops dramatically.

19.4 Your Expiration Date

As we saw in Fig. 19.1, our chance of referencing an older paper decays quickly with the paper's age. For the scientist, this prompts a critical, if uncomfortable, question: Do papers have an expiration date, after which they lose their relevance to the scientific community?

To answer this question, we can consider the average number of citations $\overline{c(t)}$ a paper acquires each year after publication. As Fig. 19.4 shows, citations follow a *jump-decay pattern* [298, 345]: A paper's citation rate rises quickly after publication, reaching a peak around year two or three, after which it starts to drop. In other words, a typical paper achieves its greatest impact within the first few years of its publication. After that, its impact diminishes quickly. As Box 19.2 explains, this jump-decay pattern is the rationale behind the two-year window used to measure the "impact factor," a metric used to evaluate journals.

The jump-decay citation pattern suggests that, if we wish to fairly compare the impact of two papers, we must first factor in their age. A paper published in 2000 may have a higher citation count than a 2010 paper simply because it has been around longer (Fig. 19.5a). Yet, as Fig. 19.5b indicates, a paper's cumulative citations do not increase indefinitely, but saturate after a few years. For example, by 2006 the citation distributions of papers published in 1991 or 1993 were statistically indistinguishable (Fig. 19.5b). In other words, the citation

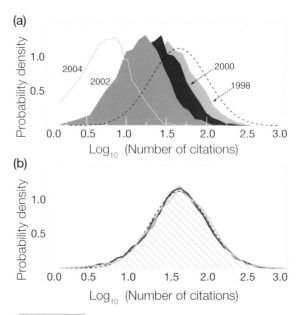

Figure 19.5 Time evolution of citation distributions. Taking papers published by *Journal of Biological Chemistry*, (a) shows the distribution of citations accrued by the end of 2006 [273]. Papers published in 2004 constitute the far-left curve, because they had only two years to collect citations. Papers published in earlier years, which have had more time to collect citations, are further to the right. However, as papers get older, this temporal shift wanes. In (b), we consider the citations to papers published between 1991 and 1993. Since those published in 1991 have mostly stopped acquiring new citations by 2006, their citation distribution is identical to that of papers published in 1993. After Stringer et al. [273].

distributions converge to a steady-state form after a certain time period (with that period varying from journal to journal) [273]. From that point on, the number of new citations a paper garners becomes negligible – those papers have reached their expiration date.

Summary

At the beginning of this chapter, we posed a difficult question: If you were at the Library of Alexandria watching the flames grow nearer, which research would you save – the old, or the new? The issues discussed in this chapter not only help us to answer this question, but may also help scientists more realistically understand the lifespan of their work.

Box 19.2 Impact factor

The impact factor (IF), often used to quantify the importance of a journal, captures the average number of citations that papers published in that journal collect within two years of publication. *Nature*, for example, has an impact factor of 38.1 for the year 2015, while *Physical Review Letters* has an impact factor of 7.6, *American Sociological Review* 4.3, and *Journal of Personality and Social Psychology* 4.7. Lately, however, the impact factor has started to take on a new, concerning function. When a paper is published in a certain journal, readers increasingly use that journal's impact factor to evaluate the potential importance of the paper. This is as misguided as trying to judge a book by its cover. Indeed, as Fig. 19.6 shows, papers published within the same journal can have vastly different impacts, telling us that the impact factor cannot predict an individual paper's impact.

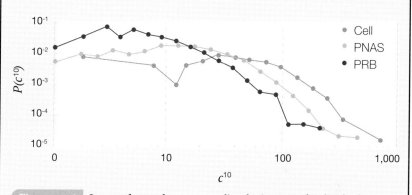

Figure 19.6 Impact factor does not predict the impact of individual papers. (a) Distribution of the cumulative citations ten years after publication (c^{10}) for all papers published in *Cell* (IF = 33.62 in 1992), *PNAS* (IF = 10.48), and *Physical Review B* (*PRB*) (IF = 3.26). Papers published by the same journal have vastly different impacts, despite sharing the same impact factor. After Wang et al. [298].

As we have discussed, scientists follow reproducible patterns when drawing upon prior knowledge. The cutting edge of new research is integral to the discovery process, given that the vast majority of references are to papers published within the last two to three years. Yet citing work from the frontier does not by itself guarantee impact: Papers with a judicious mix of new and canonical knowledge are twice as likely to be home runs than typical papers in the field. Therefore, while building upon cutting-edge work is key, a paper's impact is lifted when it considers a wide horizon of research. Put simply, if the person at the Library of Alexandria wanted to ensure that great scientific discoveries would still be possible going forward, she should try to stack up on newly published scrolls, but also grab a few older ones.

In general, a paper garners most of its citations within the first couple of years of publication. However, old papers are never entirely forgotten – hence, focusing exclusively on the impact garnered within the first few years is likely to miss the long-term impact of a discovery. That means it is only fair to compare the impact of two papers if they have had a similar amount of time to be cited, or if both papers have been around long enough that they have already collected the bulk of their citations. Often, we may not know a paper's true impact for more than a decade.

For scientists, there is another takeaway: Each of our papers have an expiration date. Whether we like it or not, there will come a point when every paper we have written stops being relevant to the community. However, this dark cloud has a silver lining: Thanks to digitization, older articles are getting more and more attention, indicating that the collective horizon of science is expanding.

But if every paper will sooner or later stop acquiring citations, what is the "ultimate" impact for each paper? And can we estimate a paper's total impact in advance?

20 ULTIMATE IMPACT

The jump-decay citation patterns described in the last chapter call to mind the notion of "15 minutes of fame." Are we left to conclude that the papers we publish – works we spend countless days and nights researching, writing, and agonizing over – will only be read for at most a few years after publication? Figure 20.1 shows the yearly citation count for 200 randomly selected papers published between 1960 and 1970 in *Physical Review*, conveying a clear message: citation patterns for individual papers are complicated. While, in aggregate citations may follow a uniform jump-decay pattern, citations of individual papers do not appear to be dictated by any apparent temporal pattern. Instead, they are remarkably variable. While most work is forgotten shortly after publication, a few papers seem to have an especially long lifespan. How can we make sense of this diverse set of trajectories?

If all papers were to follow the same jump-decay pattern, then we could easily predict the future impact of any paper after just its first few years. However, since individual discoveries take wildly different paths toward their ultimate impact, estimating the future impact of a paper appears a hopeless task. Making the matter worse is the fact that many influential discoveries are notoriously under-appreciated at first – research suggests that the more a discovery deviates from the current paradigm, the longer it will take to be appreciated by the community [59]. Indeed, given the myriad factors that affect how a new discovery is received – from the work's intrinsic value to its timing, publishing venue, and sheer chance – finding regularities in the citations of individual papers remains elusive.

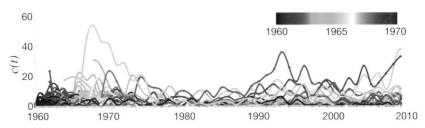

Figure 20.1 Citation histories of individual papers. We randomly selected 200 papers published between 1960 and 1970 in *Physical Review*, showing how many citations each acquired yearly following its publication. The color of the lines indicates the publication year. Blue papers were published around 1960, whereas the red were published closer to 1970. After Wang et al. [298].

Yet, as we will show in this chapter, beneath the randomness and the apparent lack of order lie some striking patterns. And these patterns make the citation dynamics of individual papers quite predictable.

20.1 Citation Dynamics of Individual Papers

Let's recap the mechanisms that we have already shown to affect the impact of a paper [298]:

- First is the *exponential growth* of science (Chapter 15). In order for papers to gain new citations, new papers must be published; hence, the rate at which these new papers are published affects how existing papers will accumulate citations.
- Second, *preferential attachment* captures the fact that highly cited papers are more visible and thus more likely than their less-cited counterparts to be cited again (Chapter 17).
- Third, *fitness* captures the inherent differences between papers, accounting for the perceived novelty and importance of a discovery (Chapter 17).
- And lastly, *aging* captures how new ideas are integrated into subsequent work: Every paper's propensity for citations eventually fades (Chapter 19), in a fashion best described by a log-normal survival probability.

What do these factors tell us about citation patterns? It turns out that we can combine these features in a mathematical model, and then solve the model analytically (see Appendix A2.4 for more detail), arriving to a

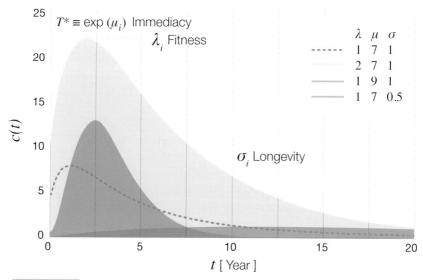

Figure 20.2 Citation patterns predicted by Eq. (20.1). The citation history of paper i is characterized by three parameters: (1) the *relative fitness* λ_i capturing a paper's ability to attract citations relative to other papers; (2) the *immediacy* μ_i captures how quickly a paper will get attention from the community, governing the time required for a paper to reach its citation peak, and (3) the *longevity* σ_i captures how quickly the attention decays over time.

formula describing the cumulative number of citations acquired by paper i at time t after publication:

$$c_i^t = m\left(e^{\lambda_i \phi\left(\frac{\ln t - \mu_i}{\sigma_i}\right)} - 1\right). \quad (20.1)$$

Here, $\phi(x) \equiv \frac{2}{\sqrt{\pi}} \int_{-\infty}^{x} e^{-y^2/2} dy$ is the cumulative normal distribution, and m corresponds to the average number of references each new paper cites. As Fig. 20.2 illustrates, for different parameters (20.1) describes a wide range of citation trajectories.

Equation (20.1) makes a remarkable prediction: As noisy and unpredictable as citation patterns of individual papers may seem (Fig. 20.1), they are all governed by the same universal equation. The differences between papers can be reduced to differences in three fundamental parameters, λ, μ, and σ. Indeed, Eq. (20.1) predicts that if we know the $(\lambda_i, \mu_i, \sigma_i)$ parameters for each paper and rescale the formula accordingly, using $\tilde{t} \equiv (\ln t - \mu_i)/\sigma_i$ and

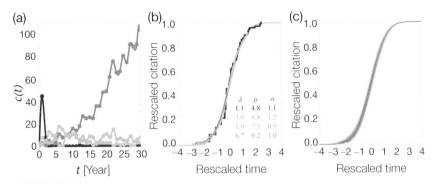

Figure 20.3 Universality in citation dynamics. (a) Citation histories of four papers published in *Physical Review* in 1964, selected for their distinct patterns: a "jump-decay" pattern (blue), a delayed peak (magenta), a constant number of citations over time (green), and an increasing number of citations each year (red). (b) Citations of an individual paper are governed by three parameters: fitness λ_i, immediacy μ_i, and longevity σ_i. Upon rescaling the citation history of each paper in (a) by the appropriate $(\lambda_i, \mu_i, \sigma_i)$ parameters, the four papers collapse into a single universal function [298]. (c) We rescaled all the papers published between 1950 and 1980 in *Physical Review* that garnered more than 30 citations in 30 years (~8,000 papers). After Wang et al. [298].

$\tilde{c} \equiv \ln\left(1 + \frac{c_i^t}{m}\right)/\lambda_i$, then each paper's citation history may follow the same universal curve:

$$\tilde{c} = \Phi(\tilde{t}). \tag{20.2}$$

In other words, all citation histories, as different as they seem, may be governed by a single formula. Next, we will test whether the empirical evidence bears out this unexpected prediction.

20.2 Citation Dynamics: Remarkably Universal

Let us select four papers whose citation histories are clearly dissimilar (Fig. 20.3a). Yet, Eq. (20.1) tells us that, if we obtain the set of $(\lambda_i, \mu_i, \sigma_i)$ parameters that best describe each paper's citation history and rescale them following (20.2), all these different curves may collapse into a single one. As Fig. 20.3b illustrates, that is precisely what happens. In other words, even though we selected these papers specifically for their diverse citation histories, none of them appears to defy this universal pattern. In fact, we could choose any number of papers published in different decades by different journals in different disciplines, as we did with some 8,000

Box 20.1 Sleeping beauties and second acts

There are some papers whose citations deviate from the typical rise-and-fall trajectories of Eq. (20.2). These come in at least two categories. First, there are "sleeping beauties," papers whose importance is not recognized until years after publication [346–348]. The classic paper by Garfield that introduced citation indices offers a fine example ([349], Fig. 20.4). Published in 1955, it only "awoke" after 2000, thanks to an increased interest in citation networks. Other sleeping beauties only awaken when they are independently discovered by a new scientific community. One example is the 1959 mathematics paper by Paul Erdős and Alfréd Rényi on random networks [350], whose impact exploded in the twenty-first century following the emergence of network science (Fig. 20.4).

The "theory of superconductivity" paper by Bardeen, Cooper, and Schrieffer (BCS) [306] offers an example of a "second act" [272], another form of atypical trajectories (Fig. 20.4). The paper took off quickly after its publication in 1957, and even won its authors the Nobel Prize in physics in 1972. Yet, soon the enthusiasm dwindled, bottoming out in 1985. However, the 1986 discovery of high-temperature superconductivity made the BCS paper relevant again, helping it experience a "second act."

These atypical trajectories are not particularly rare [347]. Indeed, in multidisciplinary journals more than 7 percent of papers can be classified as sleeping beauties. We can extend the citation model discussed in the previous section to include a second peak, helping to predict about 90 percent of the atypical cases [348].

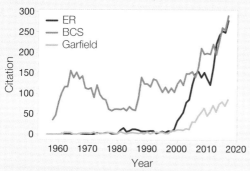

Figure 20.4 Atypical citation histories. Garfield's paper on citations is an exemplary case of a sleeping beauty [349]. Similarly, the Erdős–Rényi paper (ER) was highly regarded within mathematics but had only limited impact outside the field. However, the emergence of network science in 1999 drew new, multidisciplinary attention to the paper, fueling its explosive citation count. The superconductivity paper by Bardeen, Cooper, and Schrieffer (BCS) experienced a second act when high-temperature superconductivity was discovered.

physics papers in Fig. 20.3c, and according to Eq. (20.2) they should all collapse on the same universal curve. As our sample of papers grows, there are inevitably some that have had irregular citation histories (see also Box 20.1). Yet, remarkably, once we determine the parameters for each paper and rescale their trajectories accordingly, their citation histories largely follow the same curve (20.2).

As such, all observed differences in citation histories can be attributed to three measurable parameters: fitness, immediacy and longevity. This reveals a surprising degree of regularity in a system that seems noisy, unpredictable, and driven by countless other factors. This regularity is rooted in the fact that citations are a collective measure, reflecting the collective opinion of the scientific community on a paper's importance. Hence individual actions do not much alter a paper's path to success or oblivion – it is the collective action of countless scientists that shape their impact. As such, recognition from the community as a collective follows highly reproducible patterns and is also dominated by a small number of detectable mechanisms.

But why does this universality matter?

20.3 Ultimate Impact

While the ups and downs of a discovery's impact presents a fascinating puzzle, many scientists are interested in the net result of all these fluctuations: They want to know what a paper's cumulative impact will be when all is said and done. Equation (20.1) offers an elegant way to calculate this cumulative impact. Indeed, if we set the time period to infinity, then (20.1) represents the total number of citations a paper will ever acquire during its lifetime, or its *ultimate impact* (c^{∞}),

$$c_i^{\infty} = m(e^{\lambda_i} - 1). \tag{20.3}$$

Recall that we needed three parameters to describe the time-dependent citation count of a paper. Yet, when it comes to ultimate impact, according to (20.3), only one parameter is relevant: the relative fitness λ. It does not matter how soon the paper starts to garner attention (immediacy, μ), or how fast its appeal decays over time (longevity, σ). *Its ultimate impact is determined by its relative fitness only*, capturing the paper's importance relative to its peers.

As such, we may evaluate the long-term impact of a paper without considering the journal in which it was published. To illustrate this, we can pick papers with a comparable fitness $\lambda \approx 1$, but published in three very different journals and follow their citation histories over their first 20 years. Despite being published in journals with widely different readership and impact, the total citation counts acquired by these papers are remarkably similar (Fig. 20.5). This is precisely what (20.3) tells us: given the similar fitness λ, they should ultimately have the same impact, $c^{\infty} = 51.5$. The takeaway: Just as we don't judge a book by its cover, we can't judge a paper by its journal. *Cell* may appear to be a more coveted venue than *PNAS* – but as long as the papers have the same fitness, ultimately they are expected to acquire the same number of citations.

20.4 Future Impact

Figure 20.6 shows a typical weather forecast, the kind that often airs on TV news. It warns of a hurricane brewing in the Caribbean and predicts where and when it will hit Tennessee. This kind of detailed prediction can help residents and emergency personnel make better plans, to stock food and seek shelter.

How did scientists acquire such a predictive capability? First, massive amounts of data on past hurricane trajectories allowed them to study in minute detail the patterns characterizing these trajectories and the physical factors that drive them. Then, using that information, they built predictive models that tell us where hurricanes tend to go, given where they have already been. Can we use the same approach to predict the impact of a publication?

It turns out, we can adapt the model introduced in this chapter to predict future citations of each paper. For this we use a paper's citation history up to year T_{Train} after publication to train the model, allowing us to estimate the parameters $\lambda_i, \mu_i, \sigma_i$ for the paper. We illustrate this in Fig. 20.7a, where we use a five-year training period to predict the subsequent trajectories of three papers. The figure shows the predicted most likely citation path (red line) with an uncertainty envelope (shaded area) for each paper. Comparing our predictions to the actual citation histories, we find that two of the three papers indeed fell within the envelope. For the third paper, the model overestimated future citations. But using a longer training period ameliorates this problem. Indeed, if we increase the

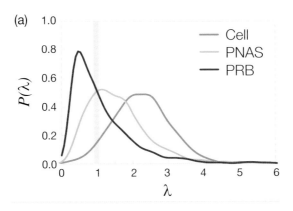
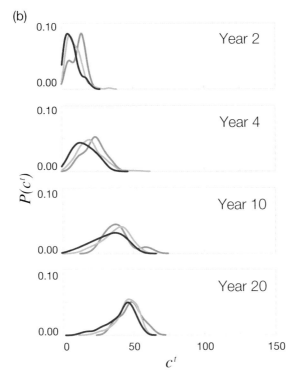

Figure 20.5 Ultimate impact. We select three journals with different readership and impact: (1) *Physical Review B (PRB)*, with an impact factor (IF) around 3.26 in 1992, is the largest journal within the *Physical Review* family, covering a specialized domain of physics; (2) *Proceedings of the National Academy of Sciences (PNAS)* (IF = 10.48) is a high-impact multidisciplinary journal covering all areas of science;

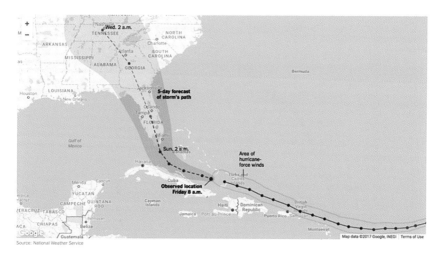

Figure 20.6 Hurricane forecast. Although the storm is thousands of miles away from Tennessee, residents know well in advance when and where it is going to hit their state and its likely intensity upon arrival. Can we develop similar predictive tools for science?

training period from 5 years to 10 years, the cone of predictions shrinks (Fig. 20.7b) and the model's predictive accuracy improves: The third paper now falls well within the prediction envelope, while the expected paths of the other two papers now align even more closely with their real-life citation trajectories. Note that the model discussed in this chapter is a minimal model, integrating mechanisms known to influence citations. As such, the examples shown in Fig. 20.7 represent an initial attempt, whose predictive accuracy and robustness can be further improved with more sophisticated models.

Figure 20.5 (cont'd) and (3) *Cell* (IF = 33.62) is an elite biology journal. For each paper published in these journals, we measure the fitness λ, obtaining their distinct journal-specific fitness distribution (a). We then select all papers with comparable fitness $\lambda \approx 1$ published in each journal and follow their citation histories. As expected, the paper's particular path depends on the journal's prestige: In early years, *Cell* papers run slightly ahead and *PRB* papers stay behind, resulting in distinct citation distributions from year 2 to year 4 ($T = 2 \div 4$). Yet, by year 20 the cumulative number of citations acquired by these papers converges in a remarkable way (b). This is precisely what (20.3) tells us: given their similar fitness λ, eventually these papers should have the same ultimate impact $c^{\infty} = 51.5$. After Wang et al. [298].

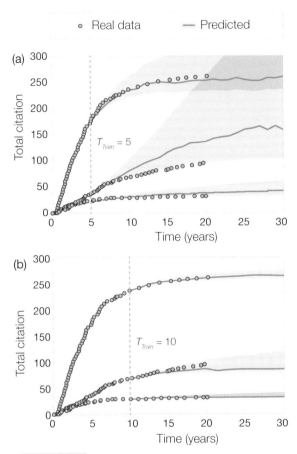

Figure 20.7 **Predicting future citations.** We can adapt the citation model to predict future citations of any given paper, by learning the model parameters from its existing citation histories. The uncertainty in estimating these parameters then translate into a prediction envelop with its most likely trajectory, similar to what we saw in the hurricane's example. As we observe more of its citation records, the envelope shrinks, and more accurately encapsulates a paper's citation history. After Wang et al. [298].

In summary, each paper follows wildly different paths on its way to achieving its ultimate impact. Yet, as we showed in this chapter, this apparent variability in citation dynamics hides a remarkable degree of regularity. Indeed, citation histories can be accurately captured by a simple model with just three parameters: fitness, immediacy, and longevity. Once we discern the three parameters characterizing a paper's citation history, we can estimate how and when the paper will accumulate its citations.

Importantly, while three parameters are required to capture the particular pattern of citation growth, when it comes to predicting the paper's ultimate impact, representing the total number of citations accumulated over a paper's lifetime, only one parameter, the fitness, matters. Indeed, as we show, papers that have the same fitness tend to acquire the same number of citations in the long run, regardless of where they are published. In contrast to impact factor and short-term citation count, which lack predictive power, ultimate impact offers a journal-independent measure of a paper's long-term impact.

The predictive accuracy of the model raises an interesting question: Could the act of predicting an article's success alter that success – or the scientific discourse at large [351]? There is reason to think so. Widespread, consistent prediction of an idea's future impact will necessarily speed up its acceptance and may undermine ideas that are predicted to fare worse, by suggesting that they do not deserve our attention or resources. As such, predicting an article's reception could act as a self-fulfilling prophecy, leading to the premature abortion of valuable ideas. On the plus side, however, the ability to better predict an article's future impact could accelerate a paper's life cycle – the time between a paper's publication and its acceptance in the community.

Part IV OUTLOOK

In the previous three parts, we aimed to offer an overview of the current body of knowledge that the science of science has offered us in the past few decades. Yet, we could not do justice to the full breadth of the subject. In this last part, we offer a brief overview of the field's emerging frontiers, discussing several areas that are just gaining traction in the community, and that have the potential to offer new, promising developments in the coming years. Unlike our approach in Parts I–III, here we do not aim to offer a comprehensive coverage, but rather to introduce some interesting problems posed by the existing research, offer some representative examples, suggest new opportunities, and imagine where they might lead us.

We will see how understanding the doings of science may change how science is done – how knowledge is discovered, hypotheses are raised, and experiments are prioritized – and what these changes may imply for individual scientists. We will consider the coming age of artificial intelligence, think through how it might impact science, and illustrate how human and machine can work together to achieve speed and efficiency that neither human nor machine can achieve alone. We will also examine how the science of science can go beyond its current attempts at correcting potential biases and how to generate causal insights with actionable policy implications.

21 CAN SCIENCE BE ACCELERATED?

In the mid eighteenth century, the steam engine jump started the Industrial Revolution, affecting most aspects of daily life and providing countries that benefitted from it with a new pathway to power and prosperity. The steam engine emerged from the somewhat counterintuitive yet profound idea that energy in one form – heat – can be converted to another – motion. While some see the steam engine as one of the most revolutionary ideas in science, it was arguably also the most overlooked [352]. Indeed, ever since we knew how to use fire to boil water, we've been quite familiar with that annoying sound the kettle lid makes as it vibrates when water reaches a rolling boil. Heat was routinely converted to motion in front of the millions, but for centuries no one seemed to recognize its practical implications.

The possibility that breakthrough ideas like the steam engine hover just under our noses, points to one of the most fruitful futures for the science of science. If our knowledge about knowledge grows in breadth and quality, will it enable researchers and decision makers to reshape our discipline, "identify areas in need of reexamination, reweight former certainties, and point out new paths that cut across revealed assumptions, heuristics, and disciplinary boundaries" [353]?

Machines have aided the scientific process for decades. Will they be able to take the next step, and help us automatically identify promising new discoveries and technologies? If so, it could drastically accelerate the advancement of science. Indeed, scientists have been relying on robot-driven laboratory instruments to screen for drugs and to sequence genomes. But, humans are still responsible for forming hypotheses, designing experiments, and drawing conclusions. What if a machine could be responsible for the *entire* scientific process – formulating a hypothesis, designing and running the experiment, analyzing data, and deciding which experiment to run next – all without human intervention? The idea may sound like a plot from a futuristic sci-fi novel, but that sci-fi scenario has, in fact, already happened. Indeed, a decade ago, back in 2009, a robotic system made a new scientific discovery with virtually no human intellectual input [354].

21.1 Close the Loop

Figure 21.1 shows "Adam," our trusty robot scientist, whose training ground was baker's yeast, an organism frequently used to model more complex life systems. While yeast is one of the best studied

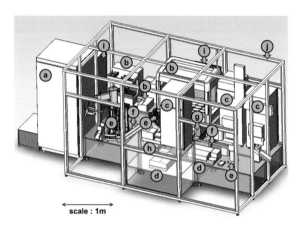

Figure 21.1 The robot scientist Adam. Adam differs from other complex laboratory systems in the individual design of the experiments to test hypotheses and the utilization of complex internal cycles. Adam has the following components: (a) an automated 20 °C freezer; (b) three liquid handlers; (c) three automated +30 °C incubators; (d) two automated plate readers; (e) three robot arms; (f) two automated plate slides; (g) an automated plate centrifuge; (h) an automated plate washer; (i) two high-efficiency particulate air filters; and (j) a rigid transparent plastic enclosure. It also has 2 bar-code readers, 7 cameras, 20 environment sensors, and 4 personal computers, as well as software, allowing it to design and initiate over 1,000 new strains and defined-growth-medium experiments each day. After King et al. [354].

organisms, the function of 10 to 15 percent of its roughly 6,000 genes remains unknown. Adam's mission was to shed light on the role of some of these mystery genes. Adam was armed with a model of yeast metabolism and a database of the genes and proteins necessary for metabolism in other species. Then it was set loose, with the role of the supervising scientists being limited to periodically add laboratory consumables and to remove waste.

Adam sought out gaps in the metabolism model, aiming to uncover "orphan" enzymes, which haven't been linked to any parent genes. After selecting a desirable orphan, Adam scoured the database for similar enzymes in other organisms, along with their corresponding genes. Using this information, Adam then hypothesized that similar genes in the yeast genome may encode the orphan enzyme and began to test this hypothesis.

It did so by performing basic operations: It selected specified yeast strains from a library held in a freezer, inoculated these strains into microtiter plate wells containing rich medium, measured their growth curves, harvested cells from each well, inoculated these cells into wells containing defined media, and measured the growth curves on the specified media. These operations are very similar to the tasks performed by a lab assistant, but they are executed robotically.

Adam is a good lab assistant, but what truly makes this machine so extraordinary is its ability to "close the loop," acting as a scientist would. After analyzing the data and running follow-up experiments, it then designed and initiated over a thousand *new* experiments. When all was said and done, Adam formulated and tested 20 hypotheses relating to genes that encode 13 different orphan enzymes. The weight of the experimental evidence for these hypotheses varied, but Adam's experiments confirmed 12 novel hypotheses.

To test Adam's findings, researchers examined the scientific literature on the 20 genes investigated. They found strong empirical evidence supporting 6 of Adam's 12 hypotheses. In other words, 6 of Adam's 12 findings were already reported in the literature, so technically they were not new. But they were new to Adam, because it had an incomplete bioinformatics database, hence was unaware that the literature has already confirmed these six hypotheses. In other words, Adam arrived at these six findings independently.

Most importantly, Adam discovered three genes which together coded for an orphan enzyme. This finding represents new knowledge that did not yet exist in the literature. And when researchers conducted the experiment by hand, they confirmed Adam's findings. The implication of this is stunning: *A machine, acting alone, created new scientific knowledge.*

We can raise some fair criticisms about Adam, particularly regarding the novelty of its findings. Although the scientific knowledge "discovered" by Adam wasn't trivial, it was implicit in the formulation of the problem, so its novelty is, arguably, modest at best. But the true value of Adam is not about what it can do *today*, but what it may be able to achieve *tomorrow*.

As a "scientist," Adam has several distinctive advantages. First, it doesn't sleep. As long as it is plugged into a power outlet, it will unceasingly, doggedly putter away in its pursuit of new knowledge. Second, this kind of "scientist" scales, easily replicable into many different copies. Third, the engine that powers Adam – including both its software and hardware – is doubling in efficiency every year. The human brain is not.

Which means Adam is only the beginning. Computers already play a key role in helping scientists store, manipulate, and analyze data. New capabilities like those offered by Adam, however, are rapidly extending the reach of computers from analysis to the formulation of hypotheses [355]. As computational tools become more powerful, they will play an increasingly important role in the genesis of scientific knowledge. They will enable automated, high-volume hypothesis generation to guide high-throughput experimentation. These experiments will likely advance a wide range of domains, from biomedicine to chemistry to physics, and even to the social sciences [355]. Indeed, as computational tools efficiently synthesize new concepts and relationships from the existing pool of knowledge, they will be able to usefully expand that pool by generating new hypotheses and drawing new conclusions [355].

These advances raise an important next question: How do we generate the most fruitful hypotheses in order to more efficiently advance science?

21.2   The Next Experiment

To improve our ability to discover new, fruitful hypotheses, we need to develop a deeper understanding of how scientists explore the knowledge frontier, and what types of exploration – or exploitation – tend to be the most fruitful. Indeed, merely increasing the pool of concepts and relationships between them typically leads to a large number of low-quality hypotheses. Instead, we need to be discerning. One example of how to home in on valuable discoveries is the Swanson hypothesis [356]. Swanson posits that if concepts A and B are studied in one literature, and B and C in another, then the link between A and C may be worth exploring. Using this approach, Swanson hypothesized that fish oil could lessen the symptoms of Raynaud's blood disorder and that magnesium deficits are linked to migraine headaches, both of which were later validated [356].

Recent attempts at applying computational tools to massive corpora of scientific texts and databases of experimental results have substantially improved our ability to trace the dynamic frontier of knowledge [319, 357]. Analyzing the abstracts of millions of biomedical papers published from 1983 to 2008, researchers identified chemicals jointly studied in a paper, represented specific research problems as links between various scientific entities, and organized them in a knowledge graph [357]. This graph allowed them to infer the typical strategy used by scientists to explore a novel chemical relationship.

For example, Fig. 21.2 shows that scientists have the tendency to explore the neighborhood of *prominent* chemicals. This paints a picture of a "crowded frontier" [358], where multiple researchers focus their investigations on a very congested neighborhood of the discoverable space, rather than exploring the space of the unknown more broadly.

While these prominent chemicals may warrant more investigations, this example suggests that there could be more optimal ways to explore the map of knowledge. For example, the model estimates that, overall, the optimal strategy for uncovering 50 percent of the graph can be nearly 10 times more efficient than the random strategy, which tests all edges with equal probability. This illustrates that a deeper understanding of how science evolves could help us accelerate the discipline's growth, allowing us to strategically choose the best next experiments.

21.3   New Challenges

"White spaces" offer both opportunity and peril. There is, of course, untapped potential in exploring fresh connections. But the

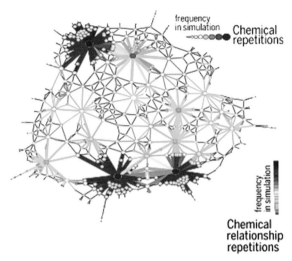

Figure 21.2 Choosing experiments to accelerate collective discovery. The actual, estimated search process illustrated on a hypothetical graph of chemical relationships. Here each node is a chemical and links indicate the pair of chemicals examined together in a paper, representing publishable chemical relationships. The graph shows simulation results averaged from 500 runs of that search strategy. The strategy swarms around a few "important," highly connected chemicals, whereas more optimal strategies estimated are much more even and less likely to "follow the crowd" in their search across the space of scientific possibilities. Adapted from [357].

underlying dangers are in the "file-drawer problem," i.e. results that never got published, which is driven by scientists' preference for publishing positive results [8, 359] and statistical findings that exceed field-specific thresholds (e.g. p-value < 0.05) [360, 361]. It is possible, therefore, that many gaps in the literature have been explored by previous generations of scientists, but, because they didn't find interpretable results, they never reported their efforts [353].

If we want to accelerate scientific progress, these negative results have positive value. Indeed, instead of being discarded, negative results should be saved, shared, compiled, and analyzed. This would not only offer a less biased view of the knowledge landscape – helping us separate opportunity from peril – it would also improve the reproducibility of findings, since we increase the credibility of positive results when we place them in the context of negative ones. This is already done in clinical trials, where results must be published at https://clinicaltrials.gov/, independent of the outcome. Indeed, biochemical journals require investigators to register early phase 1 clinical trials, as failed attempts are of critical importance to public health. For this reason, the top journals have all agreed that they won't publish the findings of phase 3 trials if their earlier phase 1 results – whether positive or negative – weren't reported first [362].

As of 2019, we're already seeing promising indications that other areas of science are moving in this direction. For example, the "preregistration revolution" encourages scientists to specify their research plans in advance, before they gather any data [363]. This practice originated in psychology, but it's being adopted by other disciplines as well. If journals and funding agencies provide the right nudges, preregistered studies may one day become the rule rather than exception.

Attempting to optimize the way we explore the knowledge landscape presents a second challenge: The design of incentive structures which encourage individual risk-taking. Whereas scientists share the overall objective of exploring the space of unknowns, individually they may hold objectives that are not optimized for science as a whole. Sociologists of science have long hypothesized that researchers' choices are shaped by an "essential tension" between productive tradition and risky innovation [364, 365]. Scientists who adhere to a research tradition often appear productive by publishing a steady stream of contributions that advance a focused research agenda. But a focused agenda may limit a researcher's ability to sense and seize opportunities for staking out new ideas. Indeed, although an innovative publication may lead to higher impact than a conservative one, high-risk innovation strategies are rare, because the potential reward does not compensate for the risk of failing to publish at all [353].

Scientific awards and accolades offer a possible way to encourage risk-taking, functioning as key incentives for taking risks to produce innovative research. Funding agencies can also help. By proactively sponsoring risky projects that test truly unexplored hypotheses, we may see more scientists venturing off well-worn research paths. But this is often easier said than done. Indeed, measurements show that the allocation of biomedical funding across topics and disciplines in the United States is better correlated to previous allocations than to the actual need, defined by the burden of diseases [366]. For example, the US spends about 55 percent of its biomedical research funding on genetic approaches, despite the fact that genome-based variation can explain only about 15–30 percent of disease causation [367]. Much less funding is devoted to environmental effects and diet, despite these being responsible for the lion's share of the disease burden experienced by the population [367], from diabetes to heart disease. These findings highlight a systemic misalignment between US health needs and research investment, casting doubt on the degree to which funding agencies – which are often run by scientists embedded in established paradigms – can successfully influence the evolution of science without additional oversight, incentives, and feedback.

To manage risk, scientists can learn from Wall Street, which invests in portfolios rather than individual stocks. While no one can really predict which experiment will work and which won't, we can spread the risk across a portfolio of experiments. Doing so would take some of the pressure off of individual research endeavors and individual scientists. For example, policymakers could design institutions that cultivate intelligent risk-taking by shifting evaluation from the individual to the group, as practiced at Bell Labs. They could also fund promising people rather than projects, an approach successfully implemented by the Howard Hughes Medical Institute. With the right incentives and institutions, researchers can choose experiments that benefit not just themselves, but also science and society more broadly [357].

*

Of course, we face challenges in realizing this vision, but the potential benefit of doing so is enormous. Eighty percent of the most transformative drugs from the past 25 years can be traced to basic scientific discoveries [368]. Shockingly, though, these discoveries were published on average *31 years* before the resulting drugs achieved FDA approval. The decades-long gap between a basic discovery to a FDA approval highlights just how vital the science of science could be in fundamentally transforming the practice of science and its policy. What if we could identify if a new discovery could lead to a new drug at time of its publication? What if we could short cut the pathway to arrive at new technologies and applications faster?

Indeed, as we discuss next, with the rise of artificial intelligence, some of these goals no longer seem far-fetched. In fact, the resulting advances may even change the very meaning of doing science.

22 ARTIFICIAL INTELLIGENCE

"What just happened?"

The sentiment circling the hallways of the CASP conference on December 2, 2018 was one of puzzlement. CASP, short for the Critical Assessment of Structure Prediction, is a biannual competition aimed at predicting the 3D structure of proteins – large, complex molecules essential for sustaining life. Predicting a protein's shape is crucial for understanding its role within a cell, as well as diagnosing and treating diseases caused by misfolded proteins, such as Alzheimer's, Parkinson's, Huntington's, and cystic fibrosis [369]. But how proteins fold their long chains of amino acids into a compact 3D shape remains one of the most important unsolved problems in biology.

Established in 1994, CASP is the Kentucky Derby of protein folding. Every two years, leading groups in the field convene to "horse-race" their best methods, establishing a new benchmark for the entire field. Then the researchers return to their labs, study each other's methods, refine and develop their own approaches, only to reconvene and race again in another two years.

At the 2018 conference, two things were unusual. First, there had been "unprecedented progress in the ability of computational methods to predict protein structure," as the organizers put it. To put that progress in perspective, roughly two competitions worth of improvement had been achieved in one. Second, this giant leap was not achieved by any of the scientists in the field. The winning team was a complete stranger to the community.

What happened at the 2018 CASP competition was merely one instance out of many in the past few years, where artificial intelligence (AI) systematically outperformed human experts in a large variety of domains. These advances have led to the consensus that the ongoing AI revolution will change almost every line of work, creating enormous social and economic opportunities, and just as many challenges [370]. As society is preparing for the moment when AI may outperform and even *replace* human doctors, drivers, soldiers, and bankers, we must ask: How will AI impact science? And, what do these changes mean for scientists?

22.1   What's New With This Wave of AI?

The technology that underlies the current AI revolution is called *deep learning*, or more specifically, deep neural networks. While there are many things AI experts have yet to agree on – including whether we should call the field "artificial intelligence" or "machine learning" – there is a consensus, in and out of academia, that this really *is* the next big thing.

A defining feature of deep learning is that it *really* works. Since 2012, it has beaten existing machine-learning techniques in more areas than we can keep track of. These advances have surely changed such typical computer science areas as image [371–374] and speech recognition [375–377], question answering [378], and language translation [379, 380]. But deep neural networks have also shattered records in such far-flung domains as predicting the activities of drugs [381], analyzing particle accelerator data [382, 383], reconstructing brain circuits [384], and predicting gene mutations and expression [385, 386].

Most importantly, many of these advances were not incremental, but represent jumps in performance. When in 2012, deep learning debuted at the ImageNet Challenge, the premier annual competition for recognizing objects in images, it almost *halved* the state-of-art error rate. Since then, deep learning algorithms are rapidly approaching human-level performance. In some cases, like strategy games such as Go or shogi (Japanese chess), and multi-player video games that emphasize collaborations, or bluffing at Texas hold'em tables, they even surpass the performance of human experts. At the CASP conference in 2018, deep learning added one more accolade to its growing resume of super-human performances: It beat all *scientists* at predicting the 3D structure of proteins.

In simple terms, AI helps us find patterns or structures in data that are implicit and probabilistic rather than explicit. These are the type of patterns which are easy for humans to find (i.e., where the cat is in a picture) but were traditionally difficult for computers. More precisely, it used to be difficult for humans to translate such tasks to computers. With AI, the machines have developed an uncanny way to do this translation themselves.

Although AI is popping up everywhere, major recent advances all hinge on a single approach: Supervised learning, where algorithms are supplied with only two sets of information – a large amount of input, or "training data," and clear instructions ("labels") for sorting the input. For example, if the goal is to identify spam emails, you supply the algorithms with millions of emails and tell it which ones are spam and which are not. The algorithm then sifts through the data to determine what kinds of emails tend to be spam. Later, when shown a new email, it tells you if it looks "spammy" based on what it has already seen.

The magic of deep learning lies in its ability to figure out the best way of representing the data without human input. That's thanks to its many intermediate layers which each offer a way to represent and transform the data based on the labels. With sufficient number of layers, the system becomes very good at discovering even the most intricate structures or patterns hidden in the data. More notably, it can discover these patterns all by itself. We can think of the different layers of deep neural networks as having the flexibility of tuning millions of knobs. As long as the system is given *enough data* with *clear directions*, it can tune all the knobs automatically and figure out the best way to represent the data.

What's so different about AI this time around? After all, more than 20 years ago, IBM's chess-playing program, Deep Blue, beat Garry Kasparov, back then the world champion. In the past AI was meticulous, but it lacked intelligence. Deep Blue defeated Mr. Kasparov because it could evaluate 200 million positions per second, allowing it to anticipate which move was most likely to lead to victory. This type of AI fails at more complex games like Go or the protein folding problem, where it can't process all the possibilities.

Deep learning, on the other hand, has been wildly successful in these arenas. In 2016, AlphaGo, created by researchers at DeepMind, defeated the world Go champion Lee Sedol over five matches. It didn't win by evaluating every possible move. Instead, it studied games of Go completed by human players, learning what kinds of moves tend to lead to victory or defeat.

But why learn from us, when the system could learn from itself? Here is where it gets *really* interesting. Merely one year after AlphaGo's victory over humans, DeepMind introduced AlphaZero [387], which has no prior knowledge or data input beyond the rules of the game. In other words, it truly started from scratch, teaching itself by repeatedly playing against itself. AlphaZero not only mastered Go, it also mastered chess and shogi, defeating every human player and computer program.

Most importantly, because AlphaZero didn't learn from human games, it doesn't play like a human. It's more like an alien, showing a kind of intuition and insight which grandmasters had never seen before. Ke Jie, the world champion of Go, remarked that AI plays "like a god." Indeed, it didn't rely on human knowledge to discover its intricate, elegant solutions. And AlphaZero managed this feat at super-human speed: After a mere four hours of chess training and eight hours of Go training, its performance exceeded the best existing programs.

Think about these numbers again. We gave an AI algorithm the rules of humanity's richest and most studied games, left it with only the rules and a board, and let it work out strategies by itself. It started by making all sorts of stupid mistakes, like all beginners do. But, by the time you came back and checked on it later that day, it had become the best player there has ever been.

If deep learning can beat humanity at its own board games, finding previously unimagined solutions to complex problems, how will it impact science, a discipline dedicated to advancing creative innovation?

22.2 The Impact of AI on Science

There are two major avenues through which AI could affect the way we do science. One is similar to what Google has done to the Internet: AI will drastically improve access to information, optimizing the various aspects of science, from information access to the automation of many processes scientists now perform. This is the utopic version, as most scientists would welcome the automation of the routine tasks, allowing us to focus on the creative process. The other avenue will be more similar to what AlphaGo has done to games of Go: AI systems could offer high-speed, creative solutions to complex problems. In a dystopic world, AI could one day replace us, scientists, moving science forward with a speed and accuracy unimaginable to us today.

22.2.1 Organizing Information

Artificial intelligence already powers many areas of modern society. Every time you type a search query into Google, AI scours the web, guessing what you really want. When you open the Facebook app, AI determines which friend's update you see first. When you shop on Amazon, AI offers you products that you may also like, even though these items were never on your shopping list. AI is also increasingly present in our devices. When you hold up your smartphone to snap a photo, AI circles in on faces and adjusts the focus to best capture them. When you address your "personal assistant" – Siri, Alexa, or Cortana – you're relying on AI to transcribe your commands into text.

What aspects of science can be augmented by this kind of AI? To begin with, today there is more literature published than we could ever hope to keep up with. Could AI identify and personalize what papers we should read? Can it cohesively summarize the text of these papers, extracting the key findings relevant to us, creating a newsletter-style digest of the key advances in the field? These new capabilities will help researchers expand the depth and quality of knowledge they acquire, as well as help identify new research possibilities.

For decision-makers in science AI could offer a more comprehensive "horizon scanning" capability, suggesting areas for strategic investment, and identifying ideas and even assembling teams that could lead to transformative science. Publishers may also use deep learning to identify which referees to seek for a manuscript, or to automatically identify apparent flaws and inconsistencies in a manuscript, avoiding the need to bother human reviewers.

Some of these applications may seem far-fetched, especially if we hope to achieve the high level of accuracy and reliability that scientists and decision-makers require. But the reality is that, although technology has substantially re-shaped human society in the past two decades, technologies that could facilitate scientific processes have languished. If you're unconvinced, just take a look at the grant submission websites of the National Science Foundation, or at the Scholar-One manuscript systems which handle most of the site's editorial functionalities – they resemble fossil websites abandoned since the dot com boom.

22.2.2 Solving Scientific Problems

Will AI one day help us pose and solve fundamental scientific problems? By integrating diverse sources of information that no individual scientist can master, can AI systems help scientists come up with more creative and better solutions faster? Can it also suggest new hypotheses, or even new areas to explore?

We've already seen some encouraging early progress in this area. For instance, researchers applied deep learning to medical diagnosis and developed an algorithm to classify a wide range of retinal pathologies, obtaining the same degree of accuracy as human experts [388]. In another example, an AI algorithm trained to classify images of skin lesions as benign or malignant achieved the accuracy of board-certified dermatologists [389]. And in emergency rooms, deep learning can now help us to decide whether a patient's CT scan shows signs of a stroke [390]. The new algorithm flagged these telltale signals at a level of accuracy comparable to medical experts' – but, crucially, it did so 150 times faster.

And of course, there's AlphaFold, the deep learning system that filled CASP attendees with awe. In the CASP contest, competing teams were given the linear sequence of amino acids for 90 proteins. The 3D shape of these proteins was known but had not yet been published. Teams then computed how the proteins would fold. By sifting through past known protein folding patterns, the predictions by AlphaFold were, on average, more accurate than any of its 97 competitors.

These successful uses of AI technology possess the two critical ingredients for deep learning: a large amount of training data and a clear way to classify it. For example, to detect skin cancers, researchers fed the algorithm with millions of images of skin lesions, telling it which ones are benign and which are malignant. Because the algorithm didn't go through the same training as dermatologists do, it may not see the same patterns that dermatologists are trained to see. This means, the AI system may also recognize patterns that have so far eluded us.

What scientific areas would most benefit from these advances? It may be helpful to think about this question in terms of the two critical ingredients for deep learning – copious data and clear perimeters for sorting it. This suggests that scientific areas that may more directly benefit from AI technology are those that are narrow enough, so that we can provide an algorithm with clear sorting strategies, but also deep enough that, by looking at *all* the data – which no scientist could ever do – could allow the AI to arrive at new results.

But most importantly, although machines are rapidly improving their efficiency and accuracy, the most exciting future of science belongs to neither humans or machines alone, but to a strategic partnership between the two.

22.3 Artificial and Human Intelligence

Let's think again about what happened in the AlphaFold case. Scientists using a new technology but without expertise or training in the specific scientific domain, were able to outperform the entire community of experts relying on traditional technologies. This example raises an important question: what if we pair the latest technology *with* researchers' subject matter expertise?

A critical area of future science of science research concerns the integration of AI so that machines and minds can work together. Ideally, AI will broaden a scientist's perspective in a way that human collaborators can't, which has far-reaching implications for science.

A recent example comes to mind. Hoping to remedy a present-day challenge in science known as the "reproducibility crisis," researchers used deep learning to uncover patterns in the narrative of scientific papers which signal strong and weak scientific findings. In 2015, the "Reproducibility Project: Psychology" (RPP) manually tested the replicability of 100 papers from top psychology journals by using the exact procedures implemented in the original studies, finding that 61 of 100 papers failed their replication test [328]. Since then, studies in psychology, economics, finance, and medicine have reported similar cases for papers that fail to reproduce [391–394].

In response, researchers combined artificial and human intelligence to estimate replicability [395]. Using data on 96 studies that underwent rigorous manual replication tests, they trained a neural network to estimate a paper's likelihood of replicability and tested the model's generalizability on 249 out-of-sample studies. The results are fascinating: The model achieved an average area under the curve (AUC) of 0.72, indicating that its predictions were significantly better than chance. To put this result in context relative to the prognosticative information provided by expert reviewers, researchers trained a new AI model that used only reviewer metrics using the same data and training procedures, finding that the reviewer metrics-only model is significantly less accurate than the narrative-only model (an AUC of 0.68). These results suggest that the AI relies on diagnostic information not captured by expert reviewers. Indeed, although the statistics reported in the paper are typically used to evaluate its merits, the accuracy of the AI shows that the narrative text actually holds more previously unexplored explanatory power. Most importantly, combining the information from the narrative model and the reviewer metrics model – in other words, combining the insights of machines and humans – yields a new AI model with the highest accuracy (AUC = 0.74).

Analyses of the mechanisms behind the model's predictive power indicate that conspicuous factors – such as word or persuasion phrase frequencies, writing style, discipline, journal, authorship, or topics – do not explain the results. Rather, the AI system uses an intricate network of linguistic relationships to predict replicability. And while words outnumber statistics in scientific papers by orders of magnitude, a paper's text has so far been largely unexploited in the science of science. Algorithms can now take advantage of the full text of papers, to detect new patterns and weak scientific findings that human experts may miss.

This example highlights a novel, and potentially formidable, human–machine partnership. Indeed, while machines can consume and digest more information than humans can, the current AI applications all belong to the category of "narrow AI," in that they can only tackle a specifically defined problem. In this respect, current AI systems are much like washing machines. They can wash any piece of clothing you throw their way, but they wouldn't know what to do with your dishes. For that, you need to build another narrow machine called a dishwasher. Similarly, we can build AI systems that are extremely good at protein folding, but which can do almost nothing else. By contrast, humans have the ability to learn, extrapolate, and think creatively in ways that machines can't.

The physics Nobel laureate Frank Wilczek famously predicted that, in 100 years, the best physicist will be a machine. Advances like AlphFold offer credibility to this prediction. But Wilczek's prediction also simplifies a complex picture: science is not only about solving well-defined problems. The most admired scientists are often those that propose new problems and identify new areas of study. Those that realize that the tools and the knowledge have advanced to the point that new discoveries are possible to break through from fallow grounds. It took humans, therefore, to realize that time is ripe to enter these new areas, and to address the challenges they present. It took humans to realize that the data and tools have matured enough that we can move forward successfully. That is, science is not only about problem solving. It is also about intuition, and ability to spot new frontiers, courage to go there, and leadership.

AI has made fabulous advances in solving problems posed by humans. It could even formulate new hypotheses within the realm of an existing body of knowledge and paradigm. Will AI ever get to the point to detect the need for a new theory, like evolution or quantum mechanics, and pursue it doggedly? As of now, there are no indications on the horizon that AI is capable of that, and many AI experts doubt it could ever be possible [396]. Hence, for now machines do not yet claim potential ownership over the future of science. Rather, our most exciting future discoveries require a strategic partnership between humans and machines. Indeed, if we assign tasks based on the abilities of each partner, scientists working in tandem with machines will potentially increase the rate of scientific progress dramatically, mitigate human blind spots, and in the process, revolutionize the practice of science.

*

There is, however, an important "but." A major drawback with the current wave of AI is that it's a black-box. Sure, it works really well, but no one knows why – which could be a big problem, especially in science. To see why, consider Amazon's experience of using AI to pick their future employees. Since 2014, Amazon has been building computer programs to review job applicants' resumes. The company's experimental AI tool rated job candidates by one to five stars – much like shoppers do when they rate products on Amazon. At first glance, it looked to be an HR holy grail. You give it 100 resumes, and it'll spit out the top five. But soon, the company realized that its new algorithm systematically penalized female candidates. The program was trained to vet applicants by observing patterns in resumes submitted to the company over a 10-year period, most of which came from men. So, the system quickly taught itself to prefer male candidates. It penalized resumes that included the word "women" and downgraded graduates of two all-women's colleges [397].

The moral of this story is not that AI can't do its job right. After all, the system did exactly what it was trained to do. Humans asked it to look at millions of past resumes, both rejects and hires, and to use this information to spot future hires. What Amazon's debacle illustrates is that, as our tools grow in accuracy and sophistication, they'll amplify and help perpetuate whatever biases we humans already have. Which means that as the science of science advances, there will be an increased need to understand biases and causal relationships in the tools and metrics our community builds.

23 BIAS AND CAUSALITY IN SCIENCE

Science of science research relies on publications and citations as its primary data sources, an approach that has important implications.

First, the explored insights and findings are limited to ideas successful enough to merit publication. Yet, most ideas fail, sometimes spectacularly. As we lack data on failures, our current understanding of how science works has multiple blind spots. Arguably, by focusing only on successful research, we perpetuate systematic biases against failure.

Second, for most of this book, the success outcomes we've used rely on citation counts. While this bias toward citations is reflective of the current landscape of the field, it highlights the need to go beyond citations as the only "currency" of science.

Third, the data-driven nature of the science of science research indicates that most studies remain observational. While such descriptive studies can reveal strong associations between events and outcomes, to understand whether a specific event "causes" a specific outcome requires us to go beyond observational studies and systematically assess causality.

Next, we will discuss in more detail these important directions for the science of science. Let's begin with a simple question: How big of a problem is our ignorance of failure?

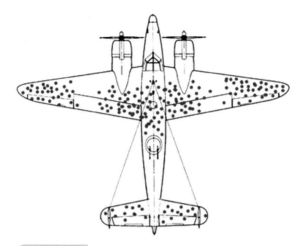

Figure 23.1 **Survivorship bias.** Illustration of hypothetical damage pattern on a WW II bomber. The damaged portions of returning planes show locations where they can take a hit and still return home safely; those hit in other places do not survive. (Image from Wikipedia under CC BY-SA 4.0.)

23.1 Failure

In World War II, the British military had access to an armor material that could make its planes bulletproof. Yet, this new armor was heavy, hence it could be only deployed to cover some, but not all, parts of a plane without compromising its flight range or maneuverability. Hence the designers faced an important question: Which part of their planes should be armored first?

The Allied forces decided to take a data-driven approach. They looked at returning B-29 bombers and they marked every place where they had taken fire. Once they gathered the data, the decision seemed simple: Just apply the armor to areas most often pocked with bullet holes (Fig. 23.1). As they moved forward with the plan, Abraham Wald, a statistician in the research group, stepped in. He explained to the Defense Department that the correct strategy should be doing exactly the opposite: applying armor to areas where bullet holes *weren't* recorded. After all, all the data comes from planes that *successfully* returned to the base. A fuselage that looks like Swiss cheese isn't a real concern if it made a round-trip journey. Instead, parts that were missing bullet holes, corresponding to engines, culprit, and other critical parts, are what need the extra protection – as those planes never made it back.

This is a wonderful example where an initial conclusion had to be completely reversed because it's based on data containing only successful samples. Similar biases abound in science: The literature tends to focus on the researchers who have successfully raised funds, published in peer-reviewed journals, patented inventions, launched new ventures, and experienced long, productive careers. These successful cases beg an important question: Given that our current understanding of science has been derived almost exclusively from successful stories, are we certain that the conclusions we've come to don't require major corrections?

Failures in science remain underexplored mainly because it's difficult to collect ground-truth information, accurately tracing failed ideas, individuals, and teams. That situation could be remedied, however, by exploring new data sources and combining them with existing ones. For example, since 2001, patent applications filed at the US Patent and Trademark Office (USPTO) have been published within 18 months of their priority date, regardless of whether they're granted or not. By tracing all applications filed, researchers can now identify successful ideas that were granted patents by the USPTO, alongside those that were turned down. Grant application databases, which contain both funded and unfunded grant proposals, represent another rich source of information for deepening our understanding of success and failure in science. A limited number of researchers have gained access to internal grant databases at funding agencies such as the US National Institutes of Health (NIH) [398], and the Netherlands Organization of Scientific Research [399]. When combined with existing publication and citation databases, this data could enable scientists to map out the rich contexts in which success and failure unfolds.

Initial investigations in this direction have offered several counterintuitive insights. Take one of our own studies as an example [398]. We focused on junior scientists whose R01 proposals to the NIH fell just above and below the funding cutoff, allowing us to compare "near-miss" with "narrow-win" individuals and to examine their longer-term career outcomes. These two groups of junior scientists were essentially the same before the "treatment" but faced a clearly different reality afterward: One group was awarded on average US$1.3 million over five years, while the other group was not. How big of a difference did this kind of early career setback make for junior scientists?

To find out, we followed both groups' subsequent career histories, finding that an early career setback is indeed of consequence: It significantly increases attrition. Indeed, one near miss predicts more than a 10 percent chance the researcher would disappear permanently from the NIH system. This rate of attrition is somewhat alarming, because to become an NIH PI, applicants have to have a demonstrated track record and years of training. A single setback, in other words, can end a career.

Yet, most surprisingly, the data indicates that the near-miss individuals who kept working as scientists systematically *outperformed* the near-winners in the long run – their publications in the next 10 years garnered significantly higher impact. This finding is quite striking. Indeed, take two researchers with similar performance who are seeking to advance their careers. The two are virtually twins, except that one has had an early funding failure and the other an early funding success. It was puzzling to discover that it's the one who *failed* who will write the higher-impact papers in the future.

One possible explanation for this finding is a screening mechanism, where the "survivors" of the near-miss group have fixed, advantageous characteristics, giving those who remain in the field better performance on average than their narrow-win counterparts. But when we accounted for this screening effect, we find that screening alone cannot explain the performance gap. In other words, those who failed but persevered didn't just start out as better performers; they also became better versions of themselves, reinforcing that "what doesn't kill you makes you stronger."

These results seem especially counterintuitive given that science is dominated by "rich get richer" dynamics, where success, not failure, brings future success. These findings therefore indicate that failure in science has powerful, opposing effects – hurting some careers, but also unexpectedly strengthening outcomes for others. Just like prior success, prior failure can also act as a marker of a future successful career. That's good news, since scientists encounter failure on a weekly if not daily basis.

More generally, though, this study raises a broader point: While we've achieved some success in understanding success, we may have failed to understand failure. Given that scientists fail more often than they succeed, understanding the whens, whys, and hows of failure, and their consequences, will not only prove essential in our attempts to understand and improve science; they may also substantially further our understanding of the human imagination by revealing the total pipeline of creative activity.

> **Box 23.1 Manuscript rejections boost the impact of the paper**
>
> A scientist who submits a new paper, only to have it rejected time after time, may conclude that the paper is simply not that good, and that even if it ever gets published, it will likely fade away into obscurity. The data shows otherwise: rejections actually boost the impact of a paper. We know this thanks to a study that tracked the submission histories of 80,748 scientific articles published in 923 bioscience journals between 2006 and 2008 [400]. The study found that resubmissions are rare: 75 percent of all papers were published in the journal to which they are first submitted. In other words, scientists are good at gauging where their papers are best suited. But when the authors compared resubmitted papers with those that made it into publications on their first try, they were in for a surprise: papers rejected on the first submission but published on the second were more highly cited within six years of publication than papers accepted immediately in the same journal. How does failure improve impact?
>
> One possibility is that authors are good at assessing the potential impact of their research. Hence, manuscripts that are initially submitted to high-impact journals are intrinsically more "fit" for citations, even if they are rejected. But this theory cannot fully explain what the researchers observed: After all, resubmissions were more highly cited regardless of whether the resubmissions moved to journals with higher or lower average impact. This suggests another possibility: Feedback from editors and reviewers, and the additional time spent revising the paper for resubmission, make for a better – and more citable – final product. So, if you've had a paper rejected, don't be frustrated with the resubmission process – what doesn't kill you may indeed make your work stronger.

23.2 A Broader Definition of Impact

Scientists have had a love–hate relationship with any metric for assessing a paper [401]. Indeed, why would researchers ever rely on proxies, rather than engage directly with the paper? And yet, citations are frequently used to gauge the recognition of a scientist by their peers. As our quantitative understanding of science improves, there is an increasing need to broaden the number and range of performance indicators. Consider the game of Go, where in each step of the game, you ought to ask yourself: What's the right move? The answer there is unambiguous – the right move is the one that will most likely win the game. Yet science lacks a single "right move," but many possible paths are interwoven. As the science of science develops, we are likely to witness the exploration of a multitude of "right moves," a better understanding of how they interact with each other, and how they capture new dimensions of scientific production and reward. These new developments will likely fall within the following three categories.

The first encompasses variants on citations. While metrics will continue to rely on the citation relationships between papers, they will likely go beyond sheer citation counts, leveraging intricate structures within the citation network. The disruption index [207], discussed in Chapter 12, offers one example: Instead of asking how many citations a paper has received, we can view each of the citations within the context of literature that's relevant to the paper. When papers that cite a given article *also* reference a substantial proportion of that article's references, then the article can be seen as consolidating the current scientific thinking. But, when the citations to an article ignore its intellectual forebears, it's an indication that the paper eclipses its prior literature, suggesting it disrupts its domain with new ideas.

The second category, alternative metrics, or altmetrics, complement the traditional citation-based impact measures. The development of Web 2.0 has changed the way research is shared within and outside academia, allowing for new innovative constructs for measuring the broader impact of a scientific work. One of the first altmetrics used for this purpose was a paper's page views. As journals moved to the Web, it is possible to count precisely how often a paper was looked at. Similarly, the discussion of a paper on various platforms can also gauge its potential impact. Scientists can calculate such metrics using data from social media, like Facebook and Twitter, blogs, and Wikipedia pages.

Researchers have calculated page views and tweets related to different papers [402, 403]. When comparing these measures with subsequent citations, there is usually a modest correlation at best. The lack of correlation with citations is both good and bad news for altmetrics. On the one hand, this lack of correlation indicates that altmetrics complement citations counts, by approximating public perception and engagement in a way that citation-based measures do not. As funders often demand a measurable outcome on the broad impact of their spending, such complementary metric could be useful. It is particularly appealing that altmetrics can be calculated shortly after a paper's publication, offering more immediate feedback than accruing citations.

On the other hand, the lack of correlation with traditional citation metrics raises questions about the usefulness of altmetrics in evaluating and predicting scientific impact. Indeed, some altmetrics are prone to self-promotion, gaming, and other mechanisms used to boost a paper's short-term visibility. After all, likes and mentions can be bought, and what's popular online may not match the value system of science. Hence, it remains unclear if and how altmetrics will be incorporated into scientific decision-making. Nevertheless, altmetrics offer a step in the right direction, helping us to diversify the way we track the impact of science beyond science. Which brings us to the third category.

An equally promising direction may be quantifying and incorporating a broader definition of impact, especially the ripples of impact that reach beyond academic science. For example, researchers analyzed the output of NIH research grants, but instead of focusing on publications and citations, they studied patents related to drugs, devices, and other medical technologies in the private sector, linking public research investments to commercial applications [404]. They found that about 10 percent of NIH grants generated a patent directly, but 30 percent generate papers that are subsequently cited in patent applications. These results indicate that academic research has a much higher impact on commercial innovation and is a lot more important than what we might have thought. In a related example, researchers measured how patented inventions built on prior scientific inquiries by mapping a network of citations between papers and patents [405]. Their research found that an astonishing 80 percent of published papers could be connected to a future patent. These two examples suggest that scientific advances and marketplace inventions are pervasively and intimately connected. By broadening their definition of impact, these studies also demonstrate the value of scientific research reaches beyond academic science. Although the vast majority of research endeavors take place in an ivory tower, that work meaningfully impacts the development of patented inventions and generates productive private-sector applications.

To be sure, more research is needed to understand what each new metric does, and how to avoid misuse. Future science of science research seeking to extend the definition of the impact will prove to be critical. For example, understanding the practical value of scientific research – particularly when it is paid for by federal funding agencies – is essential for determining how best to allocate money.

23.3   Causality

Does funding someone make that person more productive? We can answer this by collecting large-scale datasets pertaining to individual funding and linking that to subsequent productivity. Let's assume that the data shows a clear relationship between the amount of funding and the number of papers published over the next five years. Can we claim to have definitively answered our question?

Not quite. A variety of other factors both increase the likelihood of getting funding and make people appear more productive later on. For example, it could be that some fields are trendier than others, so people working in these fields are more likely to attract funding *and* also tend to publish more. Institution prestige may also play a role: It may be easier for researchers at Harvard to get funding than their colleagues at lower-tier institutions, but those Harvard professors also tend to be more productive.

We can, however, control for these factors by introducing fix effects in the regression table. The limitation of this approach is that we can only control for factors that are observable and there are many unobservable factors that are just as plausible. Take, for example, what's known as "the grit factor" [406]. Grittier researchers may be more likely to attain funding and are also more productive. These types of confounding factors are often observational studies' worst enemies, because they question whether *any* observed correlation implies a causal relationship between the two variables.

Why do we care whether the relationship is causal or not? First, understanding causality is critical for deciding what to do with the insights we glean. For example, if funding indeed results in more publications, then we can increase the amount of research in a certain area by increasing our investment in it. If, however, the observed relationship between funding and publications is entirely driven by the grit factor, then increased funding in an underexplored discipline would have little effect. Instead, we'd have to identify and support the researchers in that discipline who have grit.

Luckily, understanding causality is something science of science researchers can learn from their colleagues in economics. Over the last three decades, micro-economists have developed a series of techniques that can provide fairly reliable answers to some empirical questions. These methods, collectively known as the credibility revolution [407], rely on two basic tricks.

The first trick is utilizing a randomized controlled trial (RCT), an approach originated in medical research. To test if a certain treatment works, people participating in the trial are randomly allocated to either (1) the group receiving the treatment or (2) a control group that receives either no treatment or a placebo. The key element here is that the intervention is assigned randomly, ensuring that the treatment and control group are indistinguishable. If there are any detectable differences between the two groups after the trial, they must be the outcome of the treatment alone, and not any other factors. Let's look at an example.

We learned in Part III that teams now dominate the production of high-impact science. But we know very little about how scientific collaborations form in the first place. There is empirical evidence that new collaborations tend to occur among people who are geographically close to each other. One hypothesis posed to explain this observation is that finding collaborators presents a search problem: It's costly to find the right ones, so we tend to work with people who are in our physical proximity. That theory makes sense, but there could be alternative explanations. Maybe the geographic closeness we see between collaborators is due to the friendships cultivated in a shared workspace. Or maybe it can be explained by the shared research interests that emerge in small departments. A critical test of the search problem hypothesis would be an RCT: If we reduce search costs for some pairs of potential collaborators by facilitating face-to-face interactions but not for others – creating a treatment group and a control group – will we see any differences in collaboration?

That's precisely what researchers did when they organized a symposium for an internal funding opportunity at the Harvard Medical School [408]. They enabled face-to-face interactions during 90-minute idea-sharing sessions for some pairs of scientists who'd been randomly selected from among the 400 attendees. The probability of collaboration increased by 75 percent for the treated scientists compared with the control group, who participated in the same idea-sharing but did not have a face-to-face session. Since the two groups were randomized, the increased rate of collaboration among the treatment pairs isn't due to other factors like shared interests or pre-existing friendships. Instead, we can attribute the increased likelihood of collaboration to the fact that the pairs in question met face-to-face.

Randomized controlled trials remain the most reliable method in the world of causal inference. But experimenting on science also represents a significant challenge. Indeed, RCTs that can alter outcomes for individual researchers or institutions – which are mostly supported by tax dollars – are bound to attract criticism and push-back [79]. This is where quasi-experimental approaches will come in handy.

If we want to identify a causal relationship between factors, there's a second trick we can use: Find a randomly occurring event, use it as a cutoff, and examine what happens before and after. This is known as a "natural experiment" or a "quasi-experiment." The example on how early career setbacks lead to future success, used this trick [398]. Whether junior PIs were just above or below the funding threshold for NIH grants is an exogenous variation to any individual characteristics of the PIs themselves. People with more grit may be more resilient against setbacks, or publish higher-impact papers, but they could not choose to be in either side of this arbitrary threshold. Since being above or below the threshold only affects your chance of funding, hence if that predicts your future career outcome, it must mean that there is a link between funding and career outcome, independent of any other observable or unobservable factors.

If we can identify causal relationships in the studies we conduct, we can be more certain of the results we arrive at. Partly for this reason, the credibility revolution has had a major impact on economics. Indeed, in 2017, around 50 percent of working papers published by National Bureau of Economics Research (NBER) have the word "identification" in them [409]. But, this greater certainty comes at a price. Answers whose causal relationship can be well identified tend to cover a narrower range of topics. Indeed, one of the most critical lessons in interpreting results from RCTs is that the causal relationship, when it is established, only applies to the randomized population that was studied. A drug treatment that healed 60-year-old patients in a New York hospital won't necessarily work on adolescents in China. Similarly, effects discovered for NIH PIs around the funding threshold don't necessarily generalize to other populations. This means that describing the conditions of the experiment is just as important as describing the results. As such, RCTs and other causal identification methods highlight an essential tension between certainty and generalizability: Small, limited questions can be answered with confidence, but bigger questions are subject to much more uncertainty.

Given the tension between certainty and generalizability, both experimental and observational insights are important in our understanding of how science works. In a call for more experiments on science itself, MIT economist Pierre Azoulay pointed out [79]: "We inherited the current institutions of science from the period just after the Second World War. It would be a fortuitous coincidence if the systems that served us so well in the twentieth century were equally adapted to twenty-first-century needs." Through quasi-experiments and carefully designed experiments, the science of science will hopefully yield causal insights that will have direct policy implications for this new era.

Yet, even within economics, there is a growing consensus that many big questions in society – like the impact of global warming – cannot be answered using super-reliable techniques [410]. But, we cannot afford to ignore these big questions. Therefore, the science of science of the future will benefit from a flourishing ecology of both observational and experimental studies. By engaging in tighter partnerships with experimentalists, science of science researchers will be able to better identify associations discovered from models and large-scale data that have causal insights. Doing so will enrich their relevance in our policy and decision-making.

LAST THOUGHT: *ALL* THE SCIENCE OF SCIENCE

We would like to end this book with an invitation. We began by drawing an analogy between the recent data revolution in science and how instruments like the microscope or the telescope have transformed their disciplines. As we've discussed throughout, the science of science has advanced thanks to key contributions by researchers from a range of fields – from information and library sciences to the social, physical, and biological sciences to engineering and design – each employing distinct disciplinary models, skills, intuition, standards, and objectives to address and evaluate different aspects of the scientific enterprise.

Yet, unlike the microscope or telescope, this latest instrument has a fundamentally different characteristic. Indeed, the very nature of science of science suggests that it must be *of the sciences, by the sciences, and for the sciences*. This powerfully illustrates how a science of science that relies only on a few disciplines – the information sciences, social sciences, or engineering, let's say – will miss critical aspects of an enterprise that is growing larger, more complex, and more interconnected at an exponential rate. In other words, for science of science to succeed, it needs *all science*.

Indeed, in the next decade, the relationship between science and the science of science is expected to grow more intimate. Science of science will not just be about understanding how science is done. Instead, it will more deeply and creatively probe the question of how to do science better. This shift highlights the benefits for scientists from all disciplines to invest in science of science in order to improve their output by (1) identifying neglected yet important areas of inquiry, and

(2) doing what they already do more efficiently – by, for example, reasoning across more of the literature than can be easily read and distilling mechanisms that will guide future investigation. A tighter integration between the creation of science and our understanding of it may also enable more rapid direct and indirect replication of research, promoting reproducibility and confidence in the firmness of scientific claims.

While a broad swathe of investigators can benefit from science of science research, which is consumed by all sciences, they contribute unequally. The success of science of science depends on us overcoming traditional disciplinary barriers. These include finding publication venues that are accessible to a broad audience and which can efficiently communicate results, methods, and insights; funding sources that support a range of disciplinary perspectives and conveying insight to specific disciplines so that institutions and systems can be fine-tuned accordingly.

Embracing the science of science in its entirety will surely benefit the community of researchers engaged in furthering the field. Indeed, although several key advances in the field concern the discovery of universals across science, substantial disciplinary differences in culture, habits, and preferences make some cross-domain insights difficult to appreciate within particular fields. That can make associated policies too generic, and challenging to implement in specific areas. As science grows larger and more complex, insights from science of science that are oblivious to discipline-specific peculiarities will become increasingly inadequate.

All this means that science of science must draw upon the talents, traditions, and heuristics of every discipline if we want to make science optimally beneficial for everyone. Therefore, please accept our sincere invitation – join us in this fascinating journey.

A1 MODELING TEAM ASSEMBLY

In any team, there are two types of members: (1) Newcomers, or rookies, who have limited experience and unseasoned skills, but who often bring a refreshing, brash approach to innovation, and (2) incumbents, the veterans with proven track records, established reputations, and identifiable talents. If we categorize all scientists as either rookies or veterans, we can distinguish four different types of coauthorship links in a joint publication: (i) newcomer–newcomer, (ii) newcomer–incumbent, (iii) incumbent–incumbent, or, if both are incumbents who have worked together before, (iv) repeat incumbent–incumbent.

Varying the proportion of these four types of links within a team allows researchers to develop a model capturing how teams are assembled, which then in turn helps us understand how certain coauthorship patterns impact the team's success [187]. The proportions of newcomer–newcomer, newcomer–incumbent, incumbent–incumbent, and repeat incumbent–incumbent links can be characterized by two parameters: The *incumbency parameter*, p, which represents the fraction of incumbents within a team, and the *diversity parameter*, q, which captures the degree to which veterans involve their former collaborators.

As illustrated in Fig. A1.1, the model starts with a pool of newcomers (green circles) who haven't worked with anyone else before. Newcomers turn into incumbents (blue circles) when they are drafted onto a team for the first time. For simplicity, let's assume for

Modeling Team Assembly

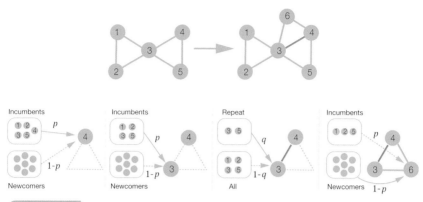

Figure A1.1 **Assembling teams in science.** The model starts with a pool of newcomers (green circles) and incumbents (blue circles). To draft team members, with a probability p we draw from the pool of incumbents, and with a probability $1 - p$, we resort to newcomers. If we decide to draw from the incumbents' pool, the diversity parameter q captures the likelihood of involving past collaborators: (i) with probability q, the new member is randomly selected from past collaborators of a randomly selected incumbent already on the team; (ii) otherwise, the new member is selected at random among all incumbents (center-right box). After Guimera et al. [187].

now that all teams have the same size. To draft team members, with probability p we draw from the pool of incumbents, and with probability $1 - p$, we resort to newcomers. If we decide to draw from the incumbents' pool and there is already another incumbent on the team (second panel), then we have another decision to make: If we need a veteran on the team, are we going to introduce a new one or bring in a past collaborator? This is captured by the diversity parameter q: (i) with probability q, the new member is randomly selected from past collaborators of a randomly selected incumbent already on the team, mimicking the tendency of existing team members to choose past collaborators; (ii) otherwise, the new member is selected at random among all incumbents (second panel).

Consider, for example, the creation of a three-person team ($m = 3$). At time zero, the collaboration network consists of five agents, all incumbents (blue circles). Along with the incumbents, there is a large pool of newcomers (green circles) eager to join new teams. As a concrete example, let us assume that incumbent 4 is selected as the first member in the new team (leftmost box). Let us also assume that the second agent is an incumbent, too (center-left box), which means we need to take the second step, to consider if we should choose from past collaborators or a new veteran. In this example, the second agent is a past collaborator of agent 4, specifically agent 3 (center-right box). Lastly, the third agent is selected from the pool of newcomers; this agent then becomes incumbent 6 (rightmost box).

The model predicts two distinct outcomes for the coauthorship network. The precise shape of the network is largely determined by the incumbency parameter p: When p is small, choosing experienced veterans is not a priority, offering ample opportunities for newcomers to enter the field. Yet frequently choosing rookies who have not worked with others before means that the coauthorship network will be fragmented into many small teams with little overlap. Increasing p increases the likelihood of having veterans on the team, thus increasing the chance that a team connects with other teams in the same network. Indeed, because veterans are on multiple teams, they are the source of overlaps throughout the network. Therefore, as p increases, the formerly fragmented teams start to form a larger cohesive cluster within the coauthorship network.

Interestingly, comparing the incumbency and diversity parameters (p and q) and a journal's impact factor, which serves as a proxy for the overall quality of the team's output, researchers find that the impact factor is positively correlated with the incumbency parameter p, but negatively correlated with the diversity parameter, q. This means that teams publishing in high-impact journals often have a higher fraction of incumbents. On the other hand, the negative correlation between the journal impact factor and diversity parameter, q, implies that teams suffer when they are composed of incumbents who mainly choose to work with prior collaborators. While these kinds of team alignments may breed familiarity, they do not breed the ingenuity that new team members can offer.

A2 MODELING CITATIONS

A2.1 The Price Model

The Price model can explain the citation disparity among scientific publications, and the universal, field-independent nature of citation distributions. To formalize the citation process, let's have m represent the number of citations on a given paper's reference list. When a scientist cites a paper, he/she does not choose it at random. Rather, the probability that the new paper cites paper i depends on how many citations i has already received, c_i:

$$\Pi_i = \frac{c_i}{\sum_i c_i}, \tag{A2.1}$$

an expression known in the network science literature as preferential attachment [304]. This means that when a scientist is choosing between two papers to cite, if one has twice as many citations as the other, she is about twice as likely to pick the more cited paper.

As written, preferential attachment (A2.1) leads to a catch-22: If a paper has no citations yet ($c_i = 0$), it can not attract new citations. We can resolve this by acknowledging that each new paper has a finite initial probability of being cited for the first time, called the initial attractiveness of a paper, c_0. Hence the probability that a paper i is cited can be modified as

$$\Pi_i = \frac{c_i + c_0}{\sum_i (c_i + c_0)}. \tag{A2.2}$$

Box A2.1 Do the rich really get richer?

Preferential attachment relies on the assumption that new papers tend to cite highly cited papers. But how do we know that preferential attachment is actually present when it comes to citations? We can answer this by measuring the citation rate of a paper (citations per year, for example) as a function of its existing citation count [20]. If preferential attachment is active, then a paper's citation rate should be linearly proportional to its total citations. Measurements have shown that this is indeed the case [272, 412, 413], offering direct empirical support for the presence of preferential attachment.

When does the rich-get-richer effect start to kick in? The answer lies in the initial attractiveness parameter, c_0, introduced in Eq. (A2.2). According to this, when a paper has very few citations ($c < c_0$), its chance of getting cited is determined mainly by the initial attractiveness, c_0. To see when a paper begins to benefit from the rich-get-richer effect, we can compare the predictions of Price's model with the Barabási–Albert model, which ignores the initial attractiveness [298], finding that the tipping point for preferential attachment is around $c_0 \approx 7$. That is, before a paper acquires 7 citations, its citations accumulate as though preferential attachment does not apply. Preferential attachment only turns on once the paper gets past this 7-citation threshold (Fig. A2.1).

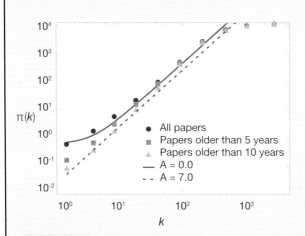

Figure A2.1 Empirical validation of initial attractiveness and preferential attachment. The solid line captures the case of initial attractiveness $c_0 = 7$ citations. The dashed line corresponds to the case without initial attractiveness ($c_0 = 0$). After Eom and Fortunato [269].

The model described above captures two key aspects of citations:

(1) **The growth of the scientific literature.** New papers are continuously published, each of which cite m previous papers.
(2) **Preferential attachment.** The probability that an author chooses a particular paper to cite is not uniform, but proportional to how many citations the paper already has.

The model with Eq. (A2.2) was first proposed by de Solla Price in 1976, hence it is sometimes called the *Price model* [297]. It allows us to analytically calculate the distribution of the number of citations received by papers, yielding:

$$p_c \sim (c + c_0)^{-\gamma}, \quad (A2.3)$$

where the citation exponent γ follows

$$\gamma = 2 + \frac{c_0}{m}. \quad (A2.4)$$

For $c \gg c_0$, (A2.3) becomes $p_c \sim c^{-\gamma}$, predicting a power-law citation distribution. Equation (A2.3) is in remarkable agreement with the citation distribution observed by Price in 1965 [262], as well as later measurements [265, 267–269, 411]. It predicts that the citation exponent (A2.4) is strictly greater than two. Many empirical measurements put the exponent around $\gamma = 3$, consistent with the case of $c_0 = m$.

A2.2 The Origins of Preferential Attachment

The rich-get-richer effect might seem to suggest that each scientist meticulously keeps track of the citation counts of every paper, so that she can cite the more cited ones. This is obviously not the case. So where does preferential attachment come from? We can answer this question by inspecting the way we encounter and cite new papers. A common way to discover the research relevant to one's work is by using the papers we have read to find other papers related to our topic of interest. That is, when we read a paper, we tend to inspect its reference list, occasionally choosing to read some of the papers referenced therein. When we later cite a paper that we discovered by reading another paper, we are effectively "copying" a citation from the earlier paper.

This copying process can help explain the origins of preferential attachment [303, 414–420]. More specifically, imagine a scientist who is deciding which papers to cite in his next paper. He could pick a random paper from the literature, something he encountered by searching for the appropriate topic or keywords. If he only chooses papers in this random way, the resulting citation distribution would follow a Poisson distribution, meaning that every citation is a random, independent event. Imagine, however, that the scientist occasionally chooses to "copy" one of the references of the paper that he randomly selected and cite that reference instead. As simple as it is, this act of copying naturally leads to preferential attachment. This is because a paper with a large number of citations will inevitably be included in multiple papers' reference lists. So, the more citations a paper has, the more likely it is that it will show up on the reference list of the paper we chose, and so the more likely it is that we will cite it.

The beauty of the copy model is that it does not require us to keep track of the citation counts of the published papers. Rather, preferential attachment arises naturally and necessarily from the model, if individuals only rely on local information (i.e., the reference lists of the papers they read previously) to find additional papers to cite. The copy model is not merely a theory – the fingerprints of this process can be detected directly in the data (Box A2.2).

A2.3 The Fit Get Richer

Price's model assumes that the growth rate of a paper's citations is determined solely by its current number of citations. To build upon this basic model, let's assume that citation rate is driven by both preferential attachment and a paper's fitness. This is called the *fitness model* or the Bianconi–Barabási model [307, 308], which incorporates the following two assumptions:

- **Growth**: In each time step, a new paper i with m references and fitness η_i is published, where η_i is a random number chosen from a distribution $p(\eta)$. Once assigned, the paper's fitness does not change over time.

Box A2.2 Evidence of citation copying

With more than 220,000 citations on Google Scholar, John P. Perdew, the pioneer of density functional theory, is one of the world's most cited physicists. His 1992 *Physical Review B* paper [421], coauthored with Y. Wang, has alone collected over 20,000 citations. However, Perdew himself has noted that thousands of those citations were likely misplaced, as many of those authors apparently intended to cite a completely different paper. Perdew and Wang had coauthored another lesser-known paper just a year before their breakthrough – but in some popular references, the paper was mistakenly listed as the more cited 1992 paper.

Analyzing such *citation misprints* can offer direct empirical evidence for citation copying [422]. Occasionally a reference to a certain paper will include a typo; for example, one digit of a paper's 4-digit page number may be misprinted. If the same misprinted number shows up repeatedly in many reference lists, this suggests that citations were simply copied from earlier publications. Indeed, the chance that multiple researchers make the same mistake independently is very low (10^{-4} in this example). Yet when researchers [422] traced a particular citation misprint back to a relatively well-known paper [423], they found that subsequent citations were disproportionately likely to carry the exact same typo. Although different authors cited the article 196 different ways, the citation with the typo was observed 78 times, suggesting that those who cited this paper may have simply copied the reference from some other paper. These repeated misprints indicate that citation "copying" is not just metaphorical, but can be quite literal.

- **Preferential attachment**: The probability that the new paper cites an existing paper i is proportional to the product of paper i's previous citations and its fitness η_i,

$$\Pi_i = \frac{\eta_i c_i}{\sum_j \eta_j c_j}. \tag{A2.5}$$

In (A2.5) the probability's dependence on c_i captures the preferential attachment mechanism we have discussed earlier. Its dependence on η_i indicates that between two papers with the same number of citations (c_i), the one with higher fitness will attract citations at a higher rate. Hence, (A2.5) assures that even a relatively new paper, with a few citations initially, can acquire citations rapidly if it has greater fitness than other papers.

A2.4 Minimal Citation Model for Individual Papers

In Part III, we discussed four different mechanisms that are shown to affect the impact of a paper [298]: the *exponential growth* of science (Chapter 15), *preferential attachment* (Chapter 17), *fitness* (Chapter 17), and *aging* (Chapter 19). Combining these four mechanisms allows us to build a minimal citation model that captures the time evolution of the citations a paper receives [298]. To do so, we write the probability that paper i is cited at time t after publication as

$$\Pi_i(t) \sim \eta_i c_i^t P_i(t). \tag{A2.6}$$

In (A2.6), η_i captures the paper's fitness, which is a collective measure capturing the community's response to the work and c_i measures preferential attachment, indicating that the paper's probability of being cited is proportional to the total number of citations it has received previously. Lastly, the long-term decay in the paper's citations is well approximated by a log-normal survival probability function

$$P_i(t) = \frac{1}{\sqrt{2\pi}\sigma_i t} \exp\left(-\frac{(\ln t - \mu_i)^2}{2\sigma_i^2}\right), \tag{A2.7}$$

where t is time elapsed since publication; μ captures the immediacy of impact, governing the time required for a paper to reach its citation peak; and σ is longevity, capturing the decay rate of citations.

The growth rate (A2.8) helps us calculate the rate at which paper i acquires new citations at time t after its publication,

$$\frac{dc_i^t}{dN} = \frac{\Pi_i}{\sum_{i=1}^{N} \Pi_i}, \tag{A2.8}$$

where N represents the total number of papers, with $N(t) \approx \exp(\beta t)$, and β characterizes the rate of science's exponential growth (Chapter 15). The rate equation (A2.8) tells us that with the publication of each new paper, paper i has a smaller and smaller chance of acquiring an additional citation. The analytical solution of the master equation (A2.8) leads to the closed-form solution, Eq. (20.1), predicting the cumulative number of citations acquired by paper i at time t after publication.

Table A2.1 shows average citation counts $<c>$ up to 2012 for all subject categories in 2004 [424]. The table also lists the number of publications, N, in each category. This table shows the remarkable heterogeneity in citations across fields, highlighting the importance of normalizing citations when comparing and evaluating impacts, as discussed in Chapter 16.

Table A2.1 The average citation counts $<c>$ up to 2012 for all subject categories in 2004 and the number of publications, N, in each category [424].

Subject categories	$<c>$	N
Physics, Acoustics	8.78	3,361
Agriculture, Agricultural Economics & Policy	7.87	592
Agriculture, Dairy & Animal Science	8.99	3,868
Agriculture, Multidisciplinary	12.07	2,803
Agriculture, Agronomy	10.0	4,767
Medicine, Allergy	18.97	1,617
Medicine, Anatomy & Morphology	11.79	1,022
Medicine, Andrology	11.46	248
Medicine, Anesthesiology	10.06	4,122
Physics, Astronomy & Astrophysics	21.41	13,392
Engineering, Automation & Control Systems	12.91	3,449
Behavioral Sciences	16.95	3,426
Biology, Biochemical Research Methods	20.54	9,674
Biology, Biochemistry & Molecular Biology	26.18	43,556
Biology, Biodiversity Conservation	14.03	2,117
Biology	16.13	5,302
Biophysics	19.48	9,609
Biology, Biotechnology & Applied Microbiology	19.63	13,899
Medicine, Cardiac & Cardiovascular Systems	20.21	12,472
Biology, Cell & Tissue Engineering	31.4	322
Biology, Cell Biology	32.72	17,610
Chemistry, Analytical	15.04	14,446
Chemistry, Applied	11.76	7,542
Chemistry, Inorganic & Nuclear	12.33	10,219
Chemistry, Medicinal	14.62	6,444
Chemistry, Multidisciplinary	21.38	23,501

Table A2.1 cont'd

Subject categories	<c>	N
Chemistry, Organic	14.56	16,878
Chemistry, Physical	18.52	29,735
Medicine, Clinical Neurology	16.95	15,563
Computer Science, Artificial Intelligence	16.77	4,690
Computer Science, Cybernetics	10.42	1,068
Computer Science, Hardware & Architecture	9.83	2,890
Computer Science, Information Systems	11.58	4,633
Computer Science, Interdisciplinary Applications	10.25	5,761
Computer Science, Software Engineering	8.85	4,718
Computer Science, Theory & Methods	9.59	3,918
Engineering, Construction & Building Technology	7.36	2,302
Medicine, Critical Care Medicine	18.15	3,116
Chemistry, Crystallography	8.1	7,032
Medicine, Oral Surgery & Medicine	11.11	5,040
Medicine, Dermatology	10.08	4,808
Biology, Developmental Biology	31.01	3,289
Ecology	18.02	9,860
Education, Scientific Disciplines	6.32	1,930
Chemistry, Electrochemistry	17.34	5,539
Medicine, Emergency Medicine	7.59	1,661
Endocrinology & Metabolism	21.68	11,259
Engineering, Energy & Fuels	11.9	5,977
Engineering, Aerospace	4.7	1,902
Engineering, Biomedical	18.82	4,717
Engineering, Chemical	10.78	13,612
Engineering, Civil	7.0	5,972
Engineering, Electrical & Electronic	11.32	26,432

Table A2.1 cont'd

Subject categories	$<c>$	N
Engineering, Environmental	16.39	4,850
Engineering, Geological	7.08	1,406
Engineering, Industrial	8.28	3,109
Engineering, Manufacturing	8.28	3,385
Engineering, Marine	1.06	489
Engineering, Mechanical	7.54	8,503
Engineering, Multidisciplinary	7.3	4,443
Engineering, Ocean	7.5	874
Engineering, Petroleum	2.14	1,613
Biology, Entomology	7.76	4,371
Environmental Sciences	14.88	16,938
Biology, Evolutionary Biology	22.88	3,170
Agriculture, Fisheries	10.79	3,495
Food Science & Technology	11.97	9,457
Forestry	11.42	2,811
Medicine, Gastroenterology & Hepatology	19.03	7,518
Biology, Genetics & Heredity	25.56	12,947
Geography, Geochemistry & Geophysics	15.79	5,777
Geography, Physical	14.6	2,230
Geography, Geology	12.42	1,604
Geography, Multidisciplinary	11.71	10,683
Geriatrics & Gerontology	15.1	2,387
Medicine, Health Care Sciences & Services	11.78	3,577
Medicine, Hematology	25.88	9,875
History & Philosophy Of Science	4.18	919
Imaging Science & Photographic Technology	16.28	1,136
Medicine, Immunology	22.17	17,048

Table A2.1 cont'd

Subject categories	$<c>$	N
Medicine, Infectious Diseases	18.47	7,727
Engineering, Instruments & Instrumentation	8.28	8,599
Medicine, Integrative & Complementary Medicine	10.45	885
Limnology	13.27	1,208
Biology, Marine & Freshwater Biology	12.33	6,939
Materials Science, Biomaterials	23.02	2,082
Materials Science, Ceramics	7.87	3,443
Materials Science, Characterization & Testing	4.59	1,293
Materials Science, Coatings & Films	11.03	4,993
Materials Science, Composites	9.44	1,539
Materials Science, Multidisciplinary	13.68	34,391
Materials Science, Paper & Wood	4.67	1,048
Materials Science, Textiles	5.72	949
Biology, Mathematical & Computational Biology	20.05	2,304
Mathematics	4.61	13,390
Mathematics, Applied	6.65	11,863
Mathematics, Interdisciplinary Applications	10.37	4,370
Mechanics	9.49	10,165
Medical Ethics	6.27	443
Medicine, Medical Informatics	11.66	1,196
Medicine, Medical Laboratory Technology	11.13	2,210
Medicine, General & Internal	19.97	14,814
Medicine, Legal	7.02	993
Medicine, Research & Experimental	20.29	8,861
Metallurgy & Metallurgical Engineering	8.12	8,077
Physics, Meteorology & Atmospheric Sciences	15.86	6,720
Biology, Microbiology	19.84	13,224

Table A2.1 cont'd

Subject categories	$<c>$	N
Microscopy	9.9	674
Engineering, Mineralogy	10.95	1,724
Engineering, Mining & Mineral Processing	6.01	1,553
Multidisciplinary Sciences	48.85	10,909
Biology, Mycology	10.45	1,019
Nanoscience & Nanotechnology	20.63	7,183
Biology, Neuroimaging	25.95	1,430
Biology, Neurosciences	23.48	23,796
Engineering, Nuclear Science & Technology	6.03	7,589
Medicine, Nursing	7.86	2,365
Medicine, Nutrition & Dietetics	18.44	4,767
Medicine, Obstetrics & Gynecology	11.78	7,384
Geography, Oceanography	13.66	4,159
Medicine, Oncology	23.44	19,647
Operations Research & Management Science	10.69	3,902
Medicine, Ophthalmology	11.97	6,359
Physics, Optics	12.16	12,693
Biology, Ornithology	8.36	928
Medicine, Orthopedics	13.0	5,607
Medicine, Otorhinolaryngology	9.01	3,235
Biology, Paleontology	9.7	1,559
Parasitology	11.06	2,239
Medicine, Pathology	14.42	5,694
Medicine, Pediatrics	10.69	9,553
Medicine, Peripheral Vascular Disease	25.41	8,353
Medicine, Pharmacology & Pharmacy	14.65	20,991
Physics, Applied	14.24	28,999

Table A2.1 cont'd

Subject categories	$<c>$	N
Physics, Atomic, Molecular & Chemical	14.32	12,973
Physics, Condensed Matter	13.39	22,654
Physics, Fluids & Plasmas	12.31	5,648
Physics, Mathematical	11.38	6,624
Physics, Multidisciplinary	16.17	15,438
Physics, Nuclear	10.24	4,987
Physics, Particles & Fields	14.33	8,759
Physiology	18.73	7,846
Biology, Plant Sciences	15.45	12,844
Materials Science, Polymer Science	14.16	11,170
Medicine, Primary Health Care	6.93	1,140
Medicine, Psychiatry	20.88	9,108
Medicine, Psychology	17.93	2,942
Medicine, Public, Environmental & Occupational Health	15.31	10,171
Medicine, Radiology, Nuclear Medicine & Medical Imaging	15.78	12,165
Rehabilitation	11.45	1,863
Engineering, Remote Sensing	14.94	1,301
Biology, Reproductive Biology	15.26	3,710
Medicine, Respiratory System	16.34	6,259
Medicine, Rheumatology	18.28	3,058
Engineering, Robotics	11.17	497
Soil Science	11.08	2,766
Materials Science, Spectroscopy	9.56	6,648
Sport Sciences	13.44	4,701
Mathematics, Statistics & Probability	9.04	4,922
Substance Abuse	16.99	1,049
Medicine, Surgery	12.25	22,687

Table A2.1 cont'd

Subject categories	$<c>$	N
Engineering, Telecommunications	9.97	5,196
Physics, Thermodynamics	9.76	3,809
Medicine, Toxicology	13.82	6,214
Medicine, Transplantation	13.21	4,665
Medicine, Transportation Science & Technology	5.44	1,562
Medicine, Tropical Medicine	11.39	1,298
Medicine, Urology & Nephrology	15.58	7,784
Medicine, Veterinary Sciences	7.64	7,967
Medicine, Virology	21.66	4,713
Water Resources	10.86	5,490
Biology, Zoology	10.07	6,684

REFERENCES

[1] W. Dennis, Bibliographies of eminent scientists. *Scientific Monthly*, **79**(3), (1954), 180–183.

[2] D. K. Simonton, Creative productivity: A predictive and explanatory model of career trajectories and landmarks. *Psychological Review*, **104**(1), (1997), 66.

[3] S. Brodetsky, Newton: Scientist and man. *Nature*, **150**, (1942), 698–699.

[4] Y. Dong, H. Ma, Z. Shen, et al., A century of science: Globalization of scientific collaborations, citations, and innovations, in *Proceedings of the 23rd ACM SIGKDD International Conference on Knowledge Discovery and Data Mining* (New York: ACM, 2017), pp. 1437–1446.

[5] R. Sinatra, P. Deville, M. Szell, et al., A century of physics. *Nature Physics*, **11**(10), (2015), 791–796.

[6] M. L. Goldberger, B. A. Maher, and P. E. E. Flattau, *Doctorate Programs in the United States: Continuity and Change* (Washington, DC: The National Academies Press, 1995).

[7] L. Baird, Departmental publication productivity and reputational quality: Disciplinary differences. *Tertiary Education and Management*, **15**(4), 2009), 355–369.

[8] J. P. Ioannidis, Why most published research findings are false. *PLoS Medicine*, **2**(8), (2005), e124.

[9] W. Shockley, On the statistics of individual variations of productivity in research laboratories. *Proceedings of the IRE*, **45**(3), (1957), 279–290.

[10] P. Fronczak, A. Fronczak, and J. A. Hołyst, Analysis of scientific productivity using maximum entropy principle and fluctuation-dissipation theorem. *Physical Review E*, **75**(2), (2007), 026103.

[11] A. J. Lotka, The frequency distribution of scientific productivity. *Journal of Washington Academy Sciences*, **16**(12), (1926), 317–324.

[12] D. de Solla Price, *Little Science, Big Science and Beyond* (New York: Columbia University Press, 1986).

[13] H. C. Lehman, Men's creative production rate at different ages and in different countries. *The Scientific Monthly*, 78, (1954), 321–326.

[14] P. D. Allison and J. A. Stewart, Productivity differences among scientists: Evidence for accumulative advantage. *American Sociological Review*, 39(4), (1974), 596–606.

[15] F. Radicchi and C. Castellano, Analysis of bibliometric indicators for individual scholars in a large data set. *Scientometrics*, 97(3), (2013), 627–637.

[16] A.-L. Barabási, *The Formula: The Universal Laws of Success* (London: Hachette, 2018).

[17] D. Bertsimas, E. Brynjolfsson, S. Reichman, et al., OR forum–tenure analytics: Models for predicting research impact. *Operations Research*, 63(6), (2015), 1246–1261.

[18] P. E. Stephan, *How Economics Shapes Science* vol. 1 (Cambridge, MA: Harvard University Press, 2012).

[19] A. Clauset, S. Arbesman, and D. B. Larremore, Systematic inequality and hierarchy in faculty hiring networks. *Science Advances*, 1(1), (2015), e1400005.

[20] W. J. Broad, The publishing game: Getting more for less. *Science*, 211 (4487), (1981), 1137–1139.

[21] N. R. Smalheiser and V. I. Torvik, Author name disambiguation. *Annual review of information science and technology*, 43(1), (2009), 1–43.

[22] A. A. Ferreira, M. A. Gonçalves, and A. H. Laender, A brief survey of automatic methods for author name disambiguation. *ACM SIGMOD Record*, 41(2), (2012), 15–26.

[23] V. I. Torvik, M. Weeber, D. R. Swanson, et al., A probabilistic similarity metric for Medline records: A model for author name disambiguation. *Journal of the American Society for Information Science and Technology*, 56(2), (2005), 140–158.

[24] A. J. Hey and P. Walters, *Einstein's Mirror* (Cambridge, UK: Cambridge University Press, 1997).

[25] D. N. Mermin, My life with Landau, in E. A. Gotsman, Y. Ne'eman, and A. Voronel, eds., *Frontiers of Physics, Proceedings of the Landau Memorial Conference* (Oxford: Pergamon Press, 1990), p. 43.

[26] J. E. Hirsch, An index to quantify an individual's scientific research output. *Proceedings of the National Academy of Sciences*, 102(46), (2005), 16569–16572.

[27] R. Van Noorden, Metrics: A profusion of measures. *Nature*, 465(7300), (2010), 864–866.

[28] A. F. Van Raan, Comparison of the Hirsch-index with standard bibliometric indicators and with peer judgment for 147 chemistry research groups. *Scientometrics*, **67**(3), (2006), 491–502.
[29] L. Zhivotovsky and K. Krutovsky, Self-citation can inflate h-index. *Scientometrics*, **77**(2), (2008), 373–375.
[30] A. Purvis, The h index: playing the numbers game. *Trends in Ecology and Evolution*, **21**(8), (2006), 422.
[31] J. E. Hirsch, Does the h index have predictive power? *Proceedings of the National Academy of Sciences*, **104**(49), (2007), 19193–19198.
[32] J. M. Cattell, *American Men Of Science: A Biographical Directory* (New York: The Science Press, 1910).
[33] J. Lane, Let's make science metrics more scientific. *Nature*, **464**(7288), (2010), 488–489.
[34] S. Alonso, F. J. Cabrerizo, E. Herrera-Viedma, et al., hg-index: A new index to characterize the scientific output of researchers based on the h-and g-indices. *Scientometrics*, **82**(2), (2009), 391–400.
[35] S. Alonso, F. J. Cabrerizo, E. Herrera-Viedma, et al., h-Index: A review focused in its variants, computation and standardization for different scientific fields. *Journal of Informetrics*, **3**(4), (2009), 273–289.
[36] Q. L. Burrell, On the h-index, the size of the Hirsch core and Jin's A-index. *Journal of Informetrics*, **1**(2), (2007), 170–177.
[37] F. J. Cabrerizo, S. Alonso, E. Herrera-Viedma, et al., q^2-Index: Quantitative and qualitative evaluation based on the number and impact of papers in the Hirsch core. *Journal of Informetrics*, **4**(1), (2010), 23–28.
[38] B. Jin, L. Liang, R. Rousseau, et al., The R-and AR-indices: Complementing the h-index. *Chinese science bulletin*, **52**(6), (2007), 855–863.
[39] M. Kosmulski, A new Hirsch-type index saves time and works equally well as the original h-index. *ISSI Newsletter*, **2**(3), (2006), 4–6.
[40] L. Egghe, An improvement of the h-index: The g-index. *ISSI newsletter*, **2**(1), (2006), 8–9.
[41] L. Egghe, Theory and practise of the g-index. *Scientometrics*, **69**(1), (2006), 131–152.
[42] S. N. Dorogovtsev and J. F. F. Mendes, Ranking scientists. *Nature Physics*, **11**(11), (2015), 882–883.
[43] F. Radicchi, S. Fortunato, and C. Castellano, Universality of citation distributions: Toward an objective measure of scientific impact. *Proceedings of the National Academy of Sciences*, **105**(45), (2008), 17268–17272.
[44] J. Kaur, F. Radicchi, and F. Menczer, Universality of scholarly impact metrics. *Journal of Informetrics*, **7**(4), (2013), 924–932.

[45] A. Sidiropoulos, D. Katsaros, and Y. Manolopoulos, Generalized Hirsch h-index for disclosing latent facts in citation networks. *Scientometrics*, 72(2), (2007), 253–280.

[46] J. Hirsch, An index to quantify an individual's scientific research output that takes into account the effect of multiple coauthorship. *Scientometrics*, 85(3), (2010), 741–754.

[47] J. E. Hirsch, h α: An index to quantify an individual's scientific leadership. *Scientometrics*, 118(2), (2019), 673–686.

[48] M. Schreiber, A modification of the h-index: The h m-index accounts for multi-authored manuscripts. *Journal of Informetrics*, 2(3), (2008), 211–216.

[49] L. Egghe, Mathematical theory of the h-and g-index in case of fractional counting of authorship. *Journal of the American Society for Information Science and Technology*, 59(10), (2008), 1608–1616.

[50] S. Galam, Tailor based allocations for multiple authorship: A fractional gh-index. *Scientometrics*, 89(1), (2011), 365.

[51] T. Tscharntke, M. E. Hochberg, T. A. Rand et al., Author sequence and credit for contributions in multiauthored publications. *PLoS Biology*, 5(1), (2007), e18.

[52] M. Ausloos, Assessing the true role of coauthors in the h-index measure of an author scientific impact. *Physica A: Statistical Mechanics and its Applications*, 422, (2015), 136–142.

[53] X. Z. Liu and H. Fang, Modifying h-index by allocating credit of multi-authored papers whose author names rank based on contribution. *Journal of Informetrics*, 6(4), (2012), 557–565.

[54] X. Hu, R. Rousseau, and J. Chen, In those fields where multiple authorship is the rule, the h-index should be supplemented by role-based h-indices. *Journal of Information Science*, 36(1), (2010), 73–85.

[55] Google Scholar. Available online at https://scholar.google.com.

[56] F. Radicchi, S. Fortunato, B. Markines, et al., Diffusion of scientific credits and the ranking of scientists. *Physical Review E*, 80(5), (2009), 056103.

[57] A. Abbott, D. Cyranoski, N. Jones, et al., Metrics: Do metrics matter? *Nature News*, 465(7300), (2010), 860–862.

[58] M. Pavlou and E. P. Diamandis, The athletes of science. *Nature*, 478(7369), (2011), 419–419.

[59] T. S. Kuhn, *The Structure of Scientific Revolutions* (Chicago: University of Chicago Press, 1962).

[60] R. K. Merton, The Matthew effect in science. *Science*, 159(3810), (1968), 56–63.

[61] T. S. Simcoe and D. M. Waguespack, Status, quality, and attention: What's in a (missing) name? *Management Science*, 57(2), (2011), 274–290.

[62] A. Tomkins, M. Zhang, and W. D. Heavlin, Reviewer bias in single-versus double-blind peer review. *Proceedings of the National Academy of Sciences*, 114(48), (2017), 12708–12713.

[63] B. McGillivray and E. De Ranieri, Uptake and outcome of manuscripts in Nature journals by review model and author characteristics. *Research Integrity and Peer Review* 3, (2018), 5, DOI: https://doi.org/10.1186/s41073-018-0049-z

[64] R. M. Blank, The effects of double-blind versus single-blind reviewing: Experimental evidence from the American Economic Review. *The American Economic Review*, 81(5), (1991), 1041–1067.

[65] A. M. Petersen, S. Fortunato, R. K. Pan, et al., Reputation and impact in academic careers. *Proceedings of the National Academy of Sciences*, 111 (2014), 15316–15321.

[66] S. Cole, Age and scientific performance. *American Journal of Sociology*, (1979), 958–977.

[67] M. Newman, *Networks: An Introduction* (Oxford: Oxford University Press, 2010).

[68] A.-L. Barabási, *Network Science* (Cambridge: Cambridge University, 2015).

[69] J. B. Fenn, M. Mann, C. K. Meng, et al., Electrospray ionization for mass spectrometry of large biomolecules. *Science*, 246(4926), (1989), 64–71.

[70] A. Mazloumian, Y.-H. Eom, D. Helbing, et al., How citation boosts promote scientific paradigm shifts and Nobel Prizes. *PloS one*, 6(5), (2011), e18975.

[71] F. C. Fang, R. G. Steen, and A. Casadevall, Misconduct accounts for the majority of retracted scientific publications. *Proceedings of the National Academy of Sciences*, 109(42), (2012), 17028–17033.

[72] S. F. Lu, G. Jin, B. Uzzi, et al., The retraction penalty: Evidence from the Web of Science. *Scientific Reports*, 3(3146), (2013).

[73] P. Azoulay, J. L. Furman, J. L. Krieger, et al., Retractions. *Review of Economics and Statistics*, 97(5), (2015), 1118–1136.

[74] P. Azoulay, A. Bonatti, and J. L. Krieger, The career effects of scandal: Evidence from scientific retractions. *Research Policy*, 46(9), (2017), 1552–1569.

[75] G. Z. Jin, B. Jones, S. Feng Lu, et al., *The Reverse Matthew Effect: Catastrophe and Consequence in Scientific Teams*, working paper 19489 (Cambridge, MA: National Bureau of Economic Research, 2013).

[76] R. K. Merton, Singletons and multiples in scientific discovery: A chapter in the sociology of science. *Proceedings of the American Philosophical Society*, 105(5), (1961), 470–486.

[77] P. Azoulay, T. Stuart, and Y. Wang, Matthew: Effect or fable? *Management Science*, 60(1), (2013), 92–109.

[78] E. Garfield, and A. Welljams-Dorof, Of Nobel class: A citation perspective on high impact research authors. *Theoretical Medicine*, 13(2), (1992), 117–135.
[79] P. Azoulay, Research efficiency: Turn the scientific method on ourselves. *Nature*, 484(7392), (2012), 31–32.
[80] M. Restivo, and A. Van De Rijt, Experimental study of informal rewards in peer production. *PloS One*, 7(3), (2012), e34358.
[81] A. van de Rijt, S. M. Kang, M. Restivo, et al., Field experiments of success-breeds-success dynamics. *Proceedings of the National Academy of Sciences*, 111(19), (2014), 6934–6939.
[82] B. Alberts, M. W. Kirschner, S. Tilghman, et al., Opinion: Addressing systemic problems in the biomedical research enterprise. *Proceedings of the National Academy of Sciences*, 112(7), (2015), 1912–1913.
[83] J. Kaiser, Biomedical research. The graying of NIH research. *Science*, 322(5903), (2008), 848–849.
[84] G. M. Beard, *Legal Responsibility in Old Age* (New York: Russells' American Steam Printing House, 1874) pp. 5–42.
[85] H. C. Lehman, *Age and Achievement* (Princeton, NJ: Princeton University Press, 1953).
[86] W. Dennis, Age and productivity among scientists. *Science*, 123, (1956), 724–725.
[87] W. Dennis, Creative productivity between the ages of 20 and 80 years. *Journal of Gerontology*, 21(1), (1966), 1–8.
[88] B. F. Jones, Age and great invention. *The Review of Economics and Statistics*, 92(1), (2010), 1–14.
[89] B. Jones, E. J. Reedy, and B. A. Weinberg, *Age and Scientific Genius*, working paper 19866 (Cambridge, MA: National Bureau of Economic Research, 2014).
[90] A. P. Usher, *A History of Mechanical Inventions*, revised edition (North Chelmsford, MA: Courier Corporation, 1954).
[91] M. L. Weitzman, Recombinant growth. *Quarterly Journal of Economics*, 113(2), (1998), 331–360.
[92] B. Uzzi, S. Mukherjee, M. Stringer, et al., Atypical combinations and scientific impact. *Science*, 342(6157), (2013), 468–472.
[93] K. A. Ericsson, R. T. Krampe, and C. Tesch-Römer, The role of deliberate practice in the acquisition of expert performance. *Psychological Review*, 100(3), (1993), 363–406.
[94] K. A. Ericsson, and A. C. Lehmann, Expert and exceptional performance: Evidence of maximal adaptation to task constraints. *Annual Review of Psychology*, 47(1), (1996), 273–305.
[95] K. A. Ericsson, R. R. Hoffman, A. Kozbelt, et al., *The Cambridge Handbook of Expertise and Expert Performance* (Cambridge, UK: Cambridge University Press, 2006).

[96] D. C. Pelz, and F. M. Andrews, *Scientists in Organizations: Productive Climates for Research and Development* (New York: Wiley, 1966).

[97] A. E. Bayer, and J. E. Dutton, Career age and research-professional activities of academic scientists: Tests of alternative nonlinear models and some implications for higher education faculty policies. *The Journal of Higher Education*, **48**(3), (1977), 259–282.

[98] R. T. Blackburn, C. E. Behymer, and D. E. Hall, Research note: Correlates of faculty publications. *Sociology of Education*, **51**(2) (1978), 132–141.

[99] K. R. Matthews, K. M. Calhoun, N. Lo, et al., The aging of biomedical research in the United States. PLoS ONE, **6**(12), (2011), e29738.

[100] C. W. Adams, The age at which scientists do their best work. Isis, **36**(3/4) (1946), 166–169.

[101] H. Zuckerman, *Scientific Elite: Nobel Laureates in the United States* (Piscataway, NJ: Transaction Publishers, 1977).

[102] D. K. Simonton, Career landmarks in science: Individual differences and interdisciplinary contrasts. *Developmental Psychology*, **27**(1), (1991), 119–130.

[103] B. F. Jones, and B. A. Weinberg, Age dynamics in scientific creativity. *Proceedings of the National Academy of Sciences*, **108**(47), (2011), 18910–18914.

[104] B. F. Jones, The burden of knowledge and the "death of the renaissance man": Is innovation getting harder? *The Review of Economic Studies*, **76**(1), (2009), 283–317.

[105] B. F. Jones, As science evolves, how can science policy? *Innovation Policy and the Economy*, **11** (2011), 103–131.

[106] F. Machlup, *The Production and Distribution of Knowledge in the United States* (Princeton, NJ: Princeton University Press, 1962).

[107] S. Fortunato, Growing time lag threatens Nobels. *Nature*, **508**(7495), (2014), 186–186.

[108] D. C. Cassidy, *Uncertainty: The Life and Science of Werner Heisenberg* (New York: Freeman, 1992), p. 1.

[109] B. A. Weinberg, and D. W. Galenson, *Creative Careers: The Life Cycles of Nobel Laureates in Economics*, working paper 11799 (Cambridge, MA: National Bureau of Economic Research, 2005).

[110] M. Rappa and K. Debackere, Youth and scientific innovation: The role of young scientists in the development of a new field. *Minerva*, **31**(1), (1993), 1–20.

[111] M. Packalen and J. Bhattacharya, *Age and the Trying Out of New Ideas*. working paper 20920 (Cambridge, MA: National Bureau of Economic Research, 2015).

[112] D. W. Galenson, *Painting Outside the Lines: Patterns of Creativity in Modern Art* (Cambridge, MA: Harvard University Press, 2009).

[113] D. W. Galenson, *Old Masters and Young Geniuses: The Two Life Cycles of Artistic Creativity* (Princeton, NJ: Princeton University Press, 2011).

[114] D. L. Hull, P. D. Tessner, and A. M. Diamond, Planck's principle. *Science*, 202(4369), (1978), 717–723.

[115] P. Azoulay, J. S. Zivin, and J. Wang, Superstar extinction. *Quarterly Journal of Economics*, 125(2), (2010), 549–589.

[116] R. Sinatra, D. Wang, P. Deville, et al., Quantifying the evolution of individual scientific impact. *Science*, 354(6312), (2016), aaf5239.

[117] L. Liu, Y. Wang, R. Sinatra, et al., Hot streaks in artistic, cultural, and scientific careers. *Nature*, 559, (2018), 396–399.

[118] D. K. Simonton, Creative productivity, age, and stress: A biographical time-series analysis of 10 classical composers. *Journal of Personality and Social Psychology*, 35(11), (1977), 791–804.

[119] D. K. Simonton, Quality, quantity, and age: The careers of ten distinguished psychologists. *International Journal of Aging & Human Development*, 21(4), (1985), 241–254.

[120] D. K. Simonton, *Genius, Creativity, and Leadership: Historiometric Inquiries* (Cambridge, MA; Harvard University Press, 1984).

[121] D. K. Simonton, *Scientific Genius: A Psychology of Science* (Cambridge, UK: Cambridge University Press, 1988).

[122] J. Li, Y. Yin, S. Fortunato, et al., Nobel laureates are almost the same as us. *Nature Reviews Physics*, 1(5), (2019), 301–303.

[123] P. Azoulay, B. F. Jones, N. J. D. Kim, et al., *Age and High-Growth Entrepreneurship* working paper 24489 (Cambridge, MA: National Bureau of Economic Research, 2018).

[124] P. Azoulay, B. Jones, J. D. King, et al., Research: The average age of a successful startup founder is 45. *Harvard Business Review*, (2018), July 11.

[125] N. Powdthavee, Y. E. Riyanto, and J. L. Knetsch, Lower-rated publications do lower academics' judgments of publication lists: Evidence from a survey experiment of economists. *Journal of Economic Psychology*, 66, (2018), 33–44.

[126] T. Gilovich, R. Vallone, and A. Tversky, The hot hand in basketball: On the misperception of random sequences. *Cognitive Psychology*, 17(3), (1985), 295–314.

[127] J. Miller and A. Sanjurjo, Surprised by the hot hand fallacy? A truth in the law of small numbers. *Econometrica*, 86(6), (2018), 2019–2047, DOI: https://doi.org/10.3982/ECTA14943.

[128] P. Ayton, and I. Fischer, The hot hand fallacy and the gambler's fallacy: Two faces of subjective randomness? *Memory & Cognition*, 32(8), (2004), 1369–1378.

[129] M. Rabin, and D. Vayanos, The gambler's and hot-hand fallacies: Theory and applications. *Review of Economic Studies*, 77(2), (2010), 730–778.

[130] J. M. Xu, and N. Harvey, Carry on winning: The gamblers' fallacy creates hot hand effects in online gambling. *Cognition*, **131**(2), (2014), 173–180.

[131] P. Csapo and M. Raab, Correction "Hand down, Man down." Analysis of defensive adjustments in response to the hot hand in basketball using novel defense metrics (vol. 9, e114184, 2014). *PLoS One*, **10**(4), (2015), e0124982.

[132] A.-L. Barabási, The origin of bursts and heavy tails in human dynamics. *Nature*, **435**(7039), (2005), 207–211.

[133] A. Vázquez, J. G. Oliveira, Z. Dezsö, et al., Modeling bursts and heavy tails in human dynamics. *Physical Review E*, **73**(3), (2006), 036127.

[134] A.-L. Barabási, *Bursts: The Hidden Patterns Behind Everything We Do, From Your E-mail to Bloody Crusades* (New York: Penguin, 2010).

[135] B. P Abbott, R. Abbott, T. D. Abbott, et al., Observation of gravitational waves from a binary black hole merger. *Physical Review Letters*, **116**(6), (2016), 061102.

[136] S. Wuchty, B.F. Jones, and B. Uzzi, The increasing dominance of teams in production of knowledge. *Science*, **316**(5827), (2007), 1036–1039.

[137] N. J. Cooke and M. L. Hilton (eds.), *Enhancing the Effectiveness of Team Science* (Washington, DC: National Academies Press, 2015).

[138] N. Drake, What is the human genome worth? *Nature News*, (2011), DOI: https://doi.org/10.1038/news.2011.281.

[139] J. Whitfield, Group theory. *Nature*, **455**(7214), (2008), 720–723.

[140] J. M. Valderas, Why do team-authored papers get cited more? *Science*, **317**(5844), (2007), 1496–1498.

[141] E. Leahey, From solo investigator to team scientist: Trends in the practice and study of research collaboration. *Annual Review of Sociology*, **42**, (2016), 81–100.

[142] C. M. Rawlings and D. A. McFarland, Influence flows in the academy: Using affiliation networks to assess peer effects among researchers. *Social Science Research*, **40**(3), (2011), 1001–1017.

[143] B. F. Jones, S. Wuchty, and B. Uzzi, Multi-university research teams: shifting impact, geography, and stratification in science. *Science*, **322**(5905), (2008), 1259–1262.

[144] Y. Xie and A. A. Killewald, *Is American Science in Decline?* (Cambridge, MA: Harvard University Press, 2012).

[145] J. Adams, Collaborations: The fourth age of research. *Nature*, **497**(7451), (2013), 557–560.

[146] Y. Xie, "Undemocracy": Inequalities in science. *Science*, **344**(6186), (2014), 809–810.

[147] M. Bikard, F. Murray, and J. S. Gans, Exploring trade-offs in the organization of scientific work: Collaboration and scientific reward. *Management Science*, **61**(7), (2015), 1473–1495.

[148] C. F. Manski, Identification of endogenous social effects: The reflection problem. *The Review of Economic Studies*, 60(3), (1993), 531–542.

[149] B. Sacerdote, Peer effects with random assignment: Results for Dartmouth roommates. *The Quarterly Journal of Economics*, 116(2), (2001), 681–704.

[150] A. Mas and E. Moretti, Peers at work. *The American Economic Review*, 99(1), (2009), 112–145.

[151] D. Herbst and A. Mas, Peer effects on worker output in the laboratory generalize to the field. *Science*, 350(6260), (2015), 545–549.

[152] A. K. Agrawal, J. McHale, and A. Oettl, *Why Stars Matter* working paper 20012 (Cambrdige, MA: National Bureau of Economic Research, 2014).

[153] J. D. Angrist and J. -S. Pischke, *Mostly Harmless Econometrics: An Empiricist's Companion* (Princeton, NJ: Princeton University Press, 2008).

[154] G. J. Borjas and K. B. Doran, Which peers matter? The relative impacts of collaborators, colleagues, and competitors. *Review of Economics and Statistics*, 97(5), (2015), 1104–1117.

[155] F. Waldinger, Peer effects in science: Evidence from the dismissal of scientists in Nazi Germany. *The Review of Economic Studies*, 79(2), (2011), 838–861.

[156] D. Crane, *Invisible Colleges: Diffusion of Knowledge in Scientific Communities* (Chicago: University of Chicago Press, 1972).

[157] A. Oettl, Sociology: Honour the helpful. *Nature*, 489(7417), (2012), 496–497.

[158] A. Oettl, Reconceptualizing stars: Scientist helpfulness and peer performance. *Management Science*, 58(6), (2012), 1122–1140.

[159] J. W. Grossman, Patterns of research in mathematics. *Notices of the AMS*, 52(1), (2005), 35–41.

[160] G. Palla, A.-L. Barabási, and T. Vicsek, Quantifying social group evolution. *Nature*, 446(7136), (2007), 664–667.

[161] J. W. Grossman and P. D. Ion, On a portion of the well-known collaboration graph. *Congressus Numerantium*, 108, (1995), 129–132.

[162] J. W. Grossman, The evolution of the mathematical research collaboration graph. *Congressus Numerantium*, 158, (2002), 201–212.

[163] A. -L Barabási, H. Jeong, Z. Neda, et al., Evolution of the social network of scientific collaborations. *Physica A: Statistical Mechanics and its Applications*, 311(3), (2002), 590–614.

[164] M. E. Newman, Coauthorship networks and patterns of scientific collaboration. *Proceedings of the National Academy of Sciences*, 101 (suppl 1), (2004), 5200–5205.

[165] M. E. Newman, The structure of scientific collaboration networks. *Proceedings of the National Academy of Sciences*, 98(2), (2001), 404–409.

[166] J. Grossman, The Erdős Number Project at Oakland University (2018), available online at https://oakland.edu/enp/.
[167] D. J. Watts and S. H. Strogatz, Collective dynamics of "small-world" networks. *Nature*, 393(6684), (1998), 440–442.
[168] B. Uzzi and J. Spiro, Collaboration and creativity: The small world Problem1. *American Journal of Sociology*, 111(2), (2005), 447–504.
[169] W. M. Muir, Group selection for adaptation to multiple-hen cages: Selection program and direct responses. *Poultry Science*, 75(4), (1996), 447–458.
[170] D. S. Wilson, *Evolution for Everyone: How Darwin's Theory Can Change the Way We Think About Our Lives* (McHenry, IL: Delta, 2007).
[171] M. A., Marks, J. E. Mathieu, and S. J. Zaccaro, A temporally based framework and taxonomy of team processes. *Academy of Management Review*, 26(3), (2001), 356–376.
[172] J. Scott, Discord turns academe's hot team cold: The self-destruction of the English department at Duke. *The New York Times*, (November 21, 1998).
[173] D. Yaffe, The department that fell to Earth: The deflation of Duke English. *Lingua Franca: The Review of Academic Life*, 9(1), (1999), 24–31.
[174] R. I. Swaab, M. Schaerer, E. M. Anicich, et al., The too-much-talent effect team interdependence determines when more talent is too much or not enough. *Psychological Science*, 25(8), (2014), 1581–1591.
[175] R. Ronay, K. Greenaway, E. M. Anicich, et al., The path to glory is paved with hierarchy when hierarchical differentiation increases group effectiveness. *Psychological Science*, 23(6), (2012), 669–677.
[176] B. Groysberg, J. T. Polzer, and H. A. Elfenbein, Too many cooks spoil the broth: How high-status individuals decrease group effectiveness. *Organization Science*, 22(3), (2011), 722–737.
[177] B. Uzzi, S. Wuchty, J. Spiro, et al., Scientific teams and networks change the face of knowledge creation, in B. Vedres and M. Scotti (eds.), *Networks in Social Policy Problems* (Cambridge: Cambridge University Press, 2012), pp. 47–59.
[178] R. B. Freeman and W. Huang, Collaboration: Strength in diversity. *Nature*, 513(7518), (2014), 305–305.
[179] R. B. Freeman and W. Huang, *Collaborating With People Like Me: Ethnic Coauthorship Within the US*, working paper 19905, (Cambridge, MA: National Bureau of Economic Research, 2014).
[180] M. J. Smith, C. Weinberger, E. M. Bruna, et al., The scientific impact of nations: Journal placement and citation performance. *PloS One*, 9(10), (2014), e109195.
[181] B. K. AlShebli, T. Rahwan, and W. L. Woon, The preeminence of ethnic diversity in scientific collaboration. *Nature Communications*, 9(1), (2018), 5163.

[182] K. Powell, These labs are remarkably diverse: Here's why they're winning at science. *Nature*, **558**(7708), (2018), 19–22.

[183] J. N. Cummings, S. Kiesler, R. Bosagh Zadeh, et al., Group heterogeneity increases the risks of large group size a longitudinal study of productivity in research groups. *Psychological Science*, **24**(6), (2013), 880–890.

[184] I. J. Deary, *Looking Down on Human Intelligence: From Psychometrics to the Brain* (Oxford: Oxford University Press, 2000).

[185] C. Spearman, "General Intelligence," objectively determined and measured. *The American Journal of Psychology*, **15**(2), (1904), 201–292.

[186] A.W. Woolley, C. F. Chabris, A. Pentland, et al., Evidence for a collective intelligence factor in the performance of human groups. *Science*, **330** (6004), (2010), 686–688.

[187] R. Guimera, B. Uzzi, J. Spiro, et al., Team assembly mechanisms determine collaboration network structure and team performance. *Science*, **308** (5722), (2005), 697–702.

[188] M. De Vaan, D. Stark, and B. Vedres, Game changer: The topology of creativity. *American Journal of Sociology*, **120**(4), (2015), 1144–1194.

[189] B. Vedres, Forbidden triads and creative success in jazz: The Miles Davis factor. *Applied Network Science*, **2**(1), (2017), 31.

[190] A. M. Petersen, Quantifying the impact of weak, strong, and super ties in scientific careers. *Proceedings of the National Academy of Sciences*, **112** (34), (2015), E4671–E4680.

[191] L. Dahlander and D. A. McFarland, Ties that last tie formation and persistence in research collaborations over time. *Administrative Science Quarterly*, **58**(1), (2013), 69–110.

[192] M. S. Brown and J. L. Goldstein, A receptor-mediated pathway for cholesterol homeostasis. *Science*, **232**(4746), (1986), 34–47.

[193] M. Heron, Deaths: Leading causes for 2012. *National Vital Statistics Reports*, **64**(10), (2015).

[194] G. Aad, B. Abbott, J. Abdallah, et al., Combined measurement of the Higgs boson mass in *pp* collisions at $\sqrt{s}$= 7 and 8 TeV with the ATLAS and CMS experiments. *Physical Review Letters*, **114**(19), (2015), 191803.

[195] D. Castelvecchi, Physics paper sets record with more than 5,000 authors. *Nature News*, May 15, 2015.

[196] S. Milojevic, Principles of scientific research team formation and evolution. *Proceedings of the National Academy of Sciences*, **111**(11), (2014), 3984–3989.

[197] M. Klug and J. P. Bagrow, Understanding the group dynamics and success of teams. *Royal Society Open Science*, **3**(4), (2016), 160007.

[198] P. B. Paulus, N. W. Kohn, L. E. Arditti, et al., Understanding the group size effect in electronic brainstorming. *Small Group Research*, **44**(3), (2013), 332–352.

[199] K. R. Lakhani, K. J. Boudreau, P.-R. Loh, et al., Prize-based contests can provide solutions to computational biology problems. *Nature Biotechnology*, **31**(2), (2013), 108–111.

[200] S. J. Barber, C. B. Harris, and S. Rajaram, Why two heads apart are better than two heads together: Multiple mechanisms underlie the collaborative inhibition effect in memory. *Journal of Experimental Psychology: Learning Memory and Cognition*, **41**(2), (2015), 559–566.

[201] J. A. Minson and J. S. Mueller, The cost of collaboration: Why joint decision-making exacerbates rejection of outside information. *Psychological Science*, **23**(3), (2012), 219–224.

[202] S. Greenstein and F. Zhu, Open content, Linus' law, and neutral point of view. *Information Systems Research*, **27**(3), (2016), 618–635.

[203] C. M. Christensen, and C. M. Christensen, *The Innovator's Dilemma: The Revolutionary Book That Will Change the Way You do Business* (New York: Harper Business Essentials, 2003).

[204] P. Bak, C. Tang, and K. Wiesenfeld, Self-organized criticality: An explanation of the 1/f noise. *Physical Review Letters*, **59**(4), (1987), 381–384.

[205] K.B. Davis, M. -O. Mewes, M. R. Andrews, et al., Bose–Einstein condensation in a gas of sodium atoms. *Physical Review Letters*, **75**(22), (1995), 3969–3973.

[206] L. Wu, D. Wang, and J. A. Evans, Large teams develop and small teams disrupt science and technology. *Nature*, **566**(7744), (2019), 378–**382.**

[207] R. J. Funk, and J. Owen-Smith, A dynamic network measure of technological change. *Management Science*, **63**(3), (2017), 791-817.

[208] A. Einstein, Die feldgleichungen der gravitation. Sitzung der physikalische-mathematischen Klasse, **25**, (1915), 844–847.

[209] J. N. Cummings and S. Kiesler, Coordination costs and project outcomes in multi-university collaborations. *Research Policy*, **36**(10), (2007), 1620–1634.

[210] M. Biagioli and P. Galison, *Scientific Authorship: Credit and Intellectual Property in Science* (Abingdon, UK: Routledge, 2014).

[211] E. A. Corrêa Jr, F. N. Silva, L. da F. Costa, et al., Patterns of authors contribution in scientific manuscripts. *Journal of Informetrics*, **11**(22), (2016), 498–510.

[212] V. Larivière, N. Desrochers, B. Macaluso, et al., Contributorship and division of labor in knowledge production. *Social Studies of Science*, **46**(3), (2016), 417–435.

[213] R. M. Slone, Coauthors' contributions to major papers published in the AJR: frequency of undeserved coauthorship. *American Journal of Roentgenology*, **167**(3), (1996), 571–579.

[214] P. Campbell, Policy on papers' contributors. *Nature*, **399**(6735), (1999), 393.

[215] V. Ilakovac, K. Fister, M. Marusic, et al., Reliability of disclosure forms of authors' contributions. *Canadian Medical Association Journal*, 176(1), (2007), 41–46.

[216] R. Deacon, M. J. Hurley, C. M. Rebolledo, et al., Nrf2: a novel therapeutic target in fragile X syndrome is modulated by NNZ2566. *Genes, Brain, and Behavior*, 16(7), (2017), 1–10.

[217] M. L. Conte, S. L. Maat, and M. B. Omary, Increased co-first authorships in biomedical and clinical publications: a call for recognition. *The FASEB Journal*, 27(10), (2013), 3902–3904.

[218] E. Dubnansky and M. B. Omary, Acknowledging joint first authors of published work: the time has come. *Gastroenterology*, 143(4), (2012), 879–880.

[219] M. B. Omary, M. B. Wallace, E. M. El-Omar, et al., A multi-journal partnership to highlight joint first-authors of manuscripts. *Gut*, 64(2), (2015), 189.

[220] D. G. Drubin, MBoC improves recognition of co-first authors. *Molecular Biology of the Cell*, 25(13), (2014), 1937.

[221] L. Waltman, An empirical analysis of the use of alphabetical authorship in scientific publishing. *Journal of Informetrics*, 6(4), (2012), 700–711.

[222] S. Jabbehdari and J. P. Walsh, Authorship norms and project structures in science. *Science, Technology, and Human Values*, 42(5), (2017), 872–900.

[223] A. G. Heffner, Authorship recognition of subordinates in collaborative research. *Social Studies of Science*, 9(3), (1979), 377–384.

[224] S. Shapin, The invisible technician. *American Scientist*, 77(6), (1989), 554–563.

[225] G. Schott, *Mechanica hydraulico-pneumatica* (Wurzburg, 1657).

[226] A. S. Rossi, Women in science: Why so few? *Science*, 148(3674), (1965), 1196–1202.

[227] Y. Xie and K. A. Shauman, *Women in Science: Career Processes and Outcomes* (Cambridge, MA: Harvard University Press, 2003).

[228] S. J. Ceci, D. K. Ginther, S. Kahn, et al., Women in academic science: A changing landscape. *Psychological Science in the Public Interest*, 15(3), (2014), 75–141.

[229] D. K. Ginther and S. Kahn, Women in economics: moving up or falling off the academic career ladder? *The Journal of Economic Perspectives*, 18(3), (2004), 193–214.

[230] H. Sarsons, Gender differences in recognition for group work. Working paper (2017), available online at https://scholar.harvard.edu/files/sarsons/files/full_v6.pdf.

[231] M. Niederle and L. Vesterlund, Do women shy away from competition? Do men compete too much? *The Quarterly Journal of Economics*, 122(3), (2007), 1067–1101.

[232] W. I. Thomas and D. S. Thomas, *The Child in America: Behavior Problems and Programs*. (New York: A. A. Knopf. 1928).
[233] R. K. Merton, The Thomas theorem and the Matthew effect. *Social Forces*, **74**(2), (1995), 379–422.
[234] R. K. Merton, *The Sociology of Science: Theoretical and Empirical Investigations* (Chicago: University of Chicago Press, 1973).
[235] G. Arnison, A. Astbury, B. Aubert, et al., Experimental observation of isolated large transverse energy electrons with associated missing energy at sqrt (s)= 540 GeV. *Physics Letters B*, **122** (1983), 103–116.
[236] H.-W. Shen and A.-L. Barabási, Collective credit allocation in science. *Proceedings of the National Academy of Sciences*, **111**(34), (2014), 12325–12330.
[237] F. Englert and R. Brout, Broken symmetry and the mass of gauge vector mesons. *Physical Review Letters*, **13**(9), (1964), 321–323.
[238] P. W. Higgs, Broken symmetries and the masses of gauge bosons. *Physical Review Letters*, **13**(16), (1964), 508–509.
[239] G. S. Guralnik, C. R. Hagen, and T. W. Kibble, Global conservation laws and massless particles. *Physical Review Letters*, **13**(20), (1964), 585–587.
[240] J.-P. Maury, *Newton: Understanding the Cosmos* (London: Thames & Hudson, 1992).
[241] D. de Solla Price, *Science Since Babylon* (New Haven, CT: Yale University Press, 1961).
[242] G. N. Gilbert and S. Woolgar, The quantitative study of science: An examination of the literature. *Science Studies*, **4**(3), (1974), 279–294.
[243] M. Khabsa and C. L. Giles, The number of scholarly documents on the public web. *PLoS One*, **9**(5), (2014), e93949.
[244] A. Sinha, Z. Shen, Y. Song, et al., An overview of Microsoft Academic Service (MAS) and applications, in *WWW '15 Companion: Proceedings of the 24th International Conference on World Wide Web* (New York: ACM, 2015), pp. 243–246.
[245] The Works of Francis Bacon *vol. IV*: Translations of the Philosophical Works ed. J. Spedding, R. L. Ellis, D. D. Heath (London: Longmans & Co., 1875), p. 109.
[246] M. Baldwin, "Keeping in the race": Physics, publication speed and national publishing strategies in *Nature*, 1895–1939. *The British Journal for the History of Science*, **47**(2), (2014), 257–279.
[247] Editorial, Form follows need. *Nature Physics*, **12**, (2016), 285.
[248] A. Csiszar, *The Scientific Journal: Authorship and the Politics of Knowledge in the Nineteenth Century* (Chicago: University of Chicago Press, 2018).
[249] C. Wendler, B. Bridgeman, F. Cline, et al., *The Path Forward: The Future of Graduate Education in the United States* (Prnceton, NJ: Educational Testing Service, 2010).

[250] Council of Graduate Schools, *PhD Completion and Attrition: Policy, Numbers, Leadership, and Next Steps* (Washington, DC: Council of Graduate Schools, 2004).
[251] M. Schillebeeckx, B. Maricque, and C. Lewis, The missing piece to changing the university culture. *Nature Biotechnology*, 31(10), (2013), 938–941.
[252] D. Cyranoski, N. Gilbert, H. Ledford, et al., Education: The PhD factory. *Nature News*, 472(7343), (2011), 276–279.
[253] Y. Yin and D. Wang, The time dimension of science: Connecting the past to the future. *Journal of Informetrics*, 11(2), (2017), 608–621.
[254] R. D. Vale, Accelerating scientific publication in biology. *Proceedings of the National Academy of Sciences*, 112(44), (2015), 13439–13446.
[255] C. Woolston, Graduate survey: A love–hurt relationship. *Nature*, 550 (7677), (2017), 549–552.
[256] K. Powell, The future of the postdoc. *Nature*, 520(7546), (2015), 144.
[257] N. Zolas, N. Goldschlag, R. Jarmin, et al., Wrapping it up in a person: Examining employment and earnings outcomes for PhD recipients. *Science*, 350(6266), (2015), 1367–1371.
[258] Editorial, Make the most of PhDs. *Nature News*, 528(7580), (2015), 7.
[259] N. Bloom, C. I. Jones, J. Van Reenen, et al., *Are Ideas Getting Harder to Find?* working paper 23782 (Cambridge, MA: National Bureau of Economic Research, 2017).
[260] S. Milojevic, Quantifying the cognitive extent of science. *Journal of Informetrics*, 9(4), (2015), 962–973.
[261] R. Van Noorden, B. Maher, and R. Nuzzo, The top 100 papers. *Nature*, 514(7524), (2014), 550–553.
[262] D. J. de Solla Price, Networks of scientific papers. *Science*, 149(3683), (1965), 510–515.
[263] E. Garfield and I. H. Sher, New factors in the evaluation of scientific literature through citation indexing. *American Documentation*, 14(3), (1963), 195–201.
[264] V. Pareto, *Cours d'économie politique* (Geneva: Librairie Droz, 1964).
[265] A. Vázquez, Statistics of citation networks. arXiv preprint https://arxiv.org/abs/cond-mat/0105031, (2001).
[266] S. Lehmann, B. Lautrup, and A. D. Jackson, Citation networks in high energy physics. *Physical Review E*, 68(2), (2003) 026113.
[267] P. O. Seglen, The skewness of science. *Journal of the American Society for Information Science*, 43(9), (1992) 628–638.
[268] I. I. Bommarito, J. Michael, and D. M. Katz, Properties of the United States code citation network. arXiv preprint https://arxiv.org/abs/0911.1751, (2009).
[269] Y.-H. Eom and S. Fortunato, Characterizing and modeling citation dynamics. *PloS One*, 6(9), (2011), e24926.

[270] F. Menczer, Evolution of document networks. *Proceedings of the National Academy of Sciences*, 101(suppl 1), (2004), 5261–5265.
[271] F. Radicchi and C. Castellano, Rescaling citations of publications in physics. *Physical Review E*, 83(4), (2011), 046116.
[272] S. Redner, Citation statistics from 110 years of *Physical Review*. *Physics Today*, 58 (2005), 49–54.
[273] M. J. Stringer, M. Sales-Pardo, and L. A. N. Amaral, Effectiveness of journal ranking schemes as a tool for locating information. *PloS One*, 3(2), (2008), e1683.
[274] C. Castellano and F. Radicchi, On the fairness of using relative indicators for comparing citation performance in different disciplines. *Archivum immunologiae et therapiae experimentalis*, 57(2), (2009), 85–90.
[275] M. J. Stringer, M. Sales-Pardo, and L. A. N. Amaral, Statistical validation of a global model for the distribution of the ultimate number of citations accrued by papers published in a scientific journal. *Journal of the American Society for Information Science and Technology*, 61(7), (2010), 1377–1385.
[276] M. L. Wallace, V. Larivière, and Y. Gingras, Modeling a century of citation distributions. *Journal of Informetrics*, 3(4), (2009), 296–303.
[277] A. D. Anastasiadis, M. P. de Albuquerque, M. P. de Albuquerque, et al., Tsallis q-exponential describes the distribution of scientific citations: A new characterization of the impact. *Scientometrics*, 83(1), (2010), 205–218.
[278] A. F. van Raan, Two-step competition process leads to quasi power-law income distributions: Application to scientific publication and citation distributions. *Physica A: Statistical Mechanics and its Applications*, 298(3), (2001), 530–536.
[279] A. F. Van Raan, Competition amongst scientists for publication status: Toward a model of scientific publication and citation distributions. *Scientometrics*, 51(1), (2001) 347–357.
[280] V. V. Kryssanov, E. L. Kuleshov, and F. J. Rinaldo et al., We cite as we communicate: A communication model for the citation process. arXiv preprint https://arxiv.org/abs/cs/0703115, (2007).
[281] A.-L. Barabási, C. Song, and D. Wang, Publishing: Handful of papers dominates citation. *Nature*, 491(7422), (2012), 40.
[282] D. W., Aksnes, Citation rates and perceptions of scientific contribution. *Journal of the American Society for Information Science and Technology*, 57(2), (2006), 169–185.
[283] F. Radicchi, In science "there is no bad publicity": Papers criticized in comments have high scientific impact. *Scientific Reports*, 2 (2012), 815.
[284] M. J. Moravcsik and P. Murugesan, Some results on the function and quality of citations. *Social Studies of Science*, 5(1), (1975), 86–92.

[285] J. R. Cole and S. Cole, *Social Stratification in Science* (Chicago: University of Chicago Press, 1973).

[286] B. Cronin, Research brief rates of return to citation. *Journal of Documentation*, **52**(2), (1996), 188–197.

[287] S. M. Lawani and A. E. Bayer, Validity of citation criteria for assessing the influence of scientific publications: New evidence with peer assessment. *Journal of the American Society for Information Science*, **34**(1), (1983), 59–66.

[288] T. Luukkonen, Citation indicators and peer review: Their time-scales, criteria of evaluation, and biases. *Research Evaluation*, **1**(1), (1991), 21–30.

[289] C. Oppenheim and S. P. Renn, Highly cited old papers and the reasons why they continue to be cited. *Journal of the American Society for Information Science*, **29**(5), (1978), 225–231.

[290] E. J. Rinia, T. N. van Leeuwen, H. G. van Vuren, et al., Comparative analysis of a set of bibliometric indicators and central peer review criteria: Evaluation of condensed matter physics in the Netherlands. *Research policy*, **27**(1), (1998), 95–107.

[291] F. Radicchi, A. Weissman, and J. Bollen, Quantifying perceived impact of scientific publications. *Journal of Informetrics*, **11**(3), (2017), 704–712.

[292] A. B. Jaffe, Patents, patent citations, and the dynamics of technological change. *NBER Reporter*, (1998, summer), 8–11.

[293] A. B. Jaffe, M. S. Fogarty, and B. A. Banks, Evidence from patents and patent citations on the impact of NASA and other federal labs on commercial innovation. *The Journal of Industrial Economics*, **46**(2), (1998), 183–205.

[294] M. Trajtenberg, A penny for your quotes: patent citations and the value of innovations. *The Rand Journal of Economics*, **221**(1), (1990), 172–187.

[295] B. H. Hall, A. B. Jaffe, and M. Trajtenberg, *Market Value and Patent Citations: A First Look*, working paper 7741, (Cambridge, MA: National Bureau of Economic Research, 2000).

[296] D. Harhoff, F. Narin, F. M. Scherer, et al., Citation frequency and the value of patented inventions. *Review of Economics and Statistics*, **81**(3), (1999), 511–515.

[297] D. de Solla Price, A general theory of bibliometric and other cumulative advantage processes. *Journal of the American Society for Information Science*, **27**(5), (1976), 292–306.

[298] D. Wang, C. Song, and A. -L. Barabási, Quantifying long-term scientific impact. *Science*, **342**(6154), (2013), 127–132.

[299] F. Eggenberger and G. Pólya, Über die statistik verketteter vorgänge. *ZAMM-Journal of Applied Mathematics and Mechanics/Zeitschrift für Angewandte Mathematik und Mechanik*, **3**(4), (1923), 279–289.

[300] G. U. Yule, A mathematical theory of evolution, based on the conclusions of Dr. JC Willis, FRS. *Philosophical Transactions of the Royal Society of London. Series B, Containing Papers of a Biological Character*, 213 (1925), 21–87.
[301] R. Gibrat, *Les inégalités économiques* (Paris: Recueil Sirey, 1931).
[302] G. K. Zipf, *Human Behavior and the Principle of Least Effort* (Boston, MA: Addison-Wesley Press, 1949).
[303] H. A. Simon, On a class of skew distribution functions. *Biometrika*, 42(3/4), (1955), 425–440.
[304] A.-L. Barabási and R. Albert, Emergence of scaling in random networks. *Science*, 286(5439), (1999), 509–512.
[305] M. E .J. Newman, The first-mover advantage in scientific publication. *EPL (Europhysics Letters)*, 86(6), (2009), 68001.
[306] J. Bardeen, L. N. Cooper, and J. R. Schrieffer, Theory of superconductivity. *Physical Review*, 108(5), (1957), 1175–1204.
[307] G. Bianconi and A. -L. Barabási, Competition and multiscaling in evolving networks. *EPL (Europhysics Letters)*, 54(4), (2001), 436–442.
[308] G. Bianconi and A. -L. Barabási, Bose–Einstein condensation in complex networks. *Physical Review Letters*,. 86(24), (2001) 5632.
[309] L. Fleming, S. Mingo, and D. Chen, Collaborative brokerage, generative creativity, and creative success. *Administrative Science Quarterly*, 52(3), (2007), 443-475.
[310] H. Youn, D. Strumsky, L. M. A. Bettencourt, et al., Invention as a combinatorial process: Evidence from US patents. *Journal of The Royal Society Interface*, 12(106), (2015), 20150272.
[311] J. Wang, R. Veugelers, and P. Stephan, Bias against novelty in science: A cautionary tale for users of bibliometric indicators. *Research Policy*, 46(8), (2017), 1416–1436.
[312] Y. -N. Lee, J. P. Walsh, and J. Wang, Creativity in scientific teams: Unpacking novelty and impact. *Research Policy*, 44(3), (2015), 684–697.
[313] C. J. Phiel, F. Zhang, E. Y. Huang, et al., Histone deacetylase is a direct target of valproic acid, a potent anticonvulsant, mood stabilizer, and teratogen. *Journal of Biological Chemistry*, 276(39), (2001), 36734–36741.
[314] P. Stephan, R. Veugelers, and J. Wang, Reviewers are blinkered by bibliometrics. *Nature News*,. 544(7651), (2017), 411.
[315] R. Van Noorden, Interdisciplinary research by the numbers. *Nature*, 525(7569), (2015), 306–307.
[316] C. S. Wagner, J. D. Roessner, K. Bobb, et al., Approaches to understanding and measuring interdisciplinary scientific research (IDR): A review of the literature. *Journal of Informetrics*, 5(1), (2011), 14–26.
[317] V. Larivière, S. Haustein, and K. Börner, Long-distance interdisciplinarity leads to higher scientific impact. *PLoS ONE*, 10(3), (2015), e0122565.

[318] E. Leahey, and J. Moody, Sociological innovation through subfield integration. *Social Currents*, 1(3), (2014), 228-256.

[319] J. G. Foster, A. Rzhetsky, and J.A. Evans, Tradition and innovation in scientists' research strategies. *American Sociological Review*, 80(5), (2015), 875–908.

[320] L. Fleming, Breakthroughs and the "long tail" of innovation. *MIT Sloan Management Review*, 49(1), (2007), 69.

[321] K. J. Boudreau, E. Guinan, K. R. Lakhani, et al., Looking across and looking beyond the knowledge frontier: Intellectual distance, novelty, and resource allocation in science. *Management Science*, 62(10), (2016), 2765–2783.

[322] L. Bromham, R. Dinnage, and X. Hua, Interdisciplinary research has consistently lower funding success. *Nature*, 534(7609), (2016), 684–687.

[323] J. Kim, C. -Y. Lee, and Y. Cho, Technological diversification, core-technology competence, and firm growth. *Research Policy*, 45(1), (2016), 113–124.

[324] D. P. Phillips, E. J. Kanter, B. Bednarczyk, et al., Importance of the lay press in the transmission of medical knowledge to the scientific community. *The New England Journal of Medicine*, 325(16), (1991), 1180–1183.

[325] F. Gonon, J. -P. Konsman, D. Cohen, et al., Why most biomedical findings echoed by newspapers turn out to be false: The case of attention deficit hyperactivity disorder. *PLoS One*, 7(9), (2012), e44275.

[326] E. Dumas-Mallet, A. Smith, T. Boraud, et al., Poor replication validity of biomedical association studies reported by newspapers. *PLoS One*, 12(2), (2017), e0172650.

[327] R. D. Peng, Reproducible research in computational science. *Science*, 334(6060), (2011), 1226–1227.

[328] Open Science Collaboration, A. Aarts, J. Anderson, et al., Estimating the reproducibility of psychological science. *Science*, 349(6251), (2015), aac4716.

[329] A. J. Wakefield, S. H. Murch, A. Anthony, et al., RETRACTED: Ileal-lymphoid-nodular hyperplasia, non-specific colitis, and pervasive developmental disorder in children. *The Lancet*, 351(1998), 637–641.

[330] C. Catalini, N. Lacetera, and A. Oettl, The incidence and role of negative citations in science. *Proceedings of the National Academy of Sciences*, 112(45), (2015), 13823–13826.

[331] F. C. Fang and A. Casadevall, Retracted science and the retraction index. *Infection and Immunity*, 79(10), (2011), 3855–3859.

[332] A. Sandison, Densities of use, and absence of obsolescence, in physics journals at MIT. *Journal of the American Society for Information Science*, 25(3), (1974), 172–182.

[333] C. Candia, C. Jara-Figueroa, C. Rodriguez-Sickert, et al., The universal decay of collective memory and attention. *Nature Human Behaviour*, 3(1), (2019), 82–91.

[334] S. Mukherjee, D. M. Romero, B. Jones, et al., The age of past knowledge and tomorrow's scientific and technological breakthroughs. *Science Advances*, 3(4), (2017), e1601315.

[335] A. Odlyzko, The rapid evolution of scholarly communication. *Learned Publishing*, 15(1), (2002), 7–19.

[336] V. Larivière, É. Archambault, and Y. Gingras, Long-term variations in the aging of scientific literature: From exponential growth to steady-state science (1900–2004). *Journal of the American Society for Information Science and Technology*, 59(2), (2008), 288–296.

[337] A. Verstak, A. Acharya H. Suzuki et al., On the shoulders of giants: The growing impact of older articles. arXiv preprint https://arxiv.org/abs/1411.0275, (2014).

[338] J. A. Evans, Electronic publication and the narrowing of science and scholarship. *Science*, 321(5887), (2008), 395–399.

[339] J. C. Burnham, The evolution of editorial peer review. *Journal of the American Medical Association*, 263(10), (1990), 1323–1329.

[340] R. Spier, The history of the peer-review process. *Trends in Biotechnology*, 20(8), (2002), 357–358.

[341] Q. L. Burrell, Modelling citation age data: Simple graphical methods from reliability theory. *Scientometrics*, 55(2), (2002), 273–285.

[342] W. Glänzel, Towards a model for diachronous and synchronous citation analyses. *Scientometrics*, 60(3), (2004), 511–522.

[343] H. Nakamoto, Synchronous and diachronous citation distribution, in L. Egghe and R. Rousseau (eds.), *Informetrics 87/88: Select Proceedings of the First International Conference on Bibliometrics and Theoretical Aspects of Information Retrieval* (Amsterdam: Elsevier Science Publishers, 1988).

[344] R. K. Pan, A. M. Petersena, F. Pammolli, et al., The memory of science: Inflation, myopia, and the knowledge network. *Journal of Informetrics*, 12, (2016), 656–678.

[345] P. D. B. Parolo, R. K. Pan, R. Ghosh, et al., Attention decay in science. *Journal of Informetrics*, 9(4), (2015), 734–745.

[346] A. F. Van Raan, Sleeping beauties in science. *Scientometrics*, 59(3), (2004), 467–472.

[347] Q. Ke, E. Ferrara, F. Radicchi, et al., Defining and identifying Sleeping Beauties in science. *Proceedings of the National Academy of Sciences*, 112(24), (2015), 7426–7431.

[348] Z. He, Z. Lei, and D. Wang, Modeling citation dynamics of "atypical" articles. *Journal of the Association for Information Science and Technology*, 69(9), (2018), 1148–1160.

[349] E. Garfield, Citation indexes for science. *Science*, **122**(3159), (1955), 108–111.
[350] P. Erdős, and A. Rényi, On the evolution of random graphs. *Publications of the Mathematical Institute of the Hungarian Academy of Sciences*, **5**(1), (1960), 17–61.
[351] J. A. Evans, Future science. *Science*, **342**(44), (2013), 44–45.
[352] Y. N. Harari, *Sapiens: A Brief History of Humankind* (London: Random House, 2014).
[353] J. A. Evans, and J. G. Foster, Metaknowledge. *Science*, **331**(6018), (2011), 721–725.
[354] R. D. King, J. Rowland, S. G. Olive et al., The automation of science. *Science*, **324**(5923), (2009), 85–89.
[355] J. Evans, and A. Rzhetsky, Machine science. *Science*, **329**(5990), (2010), 399–400.
[356] D. R. Swanson, Migraine and magnesium: Eleven neglected connections. *Perspectives in Biology and Medicine*, **31**(4), (1988), 526–557.
[357] A. Rzhetsky, J. G. Foster, I. T. Foster, et al., Choosing experiments to accelerate collective discovery. *Proceedings of the National Academy of Sciences*, **112**(47), (2015), 14569–14574.
[358] P. Azoulay, J. Graff-Zivin, B. Uzzi, et al., Toward a more scientific science. *Science*, **361**(6408), (2018), 1194–1197.
[359] S. A., Greenberg, How citation distortions create unfounded authority: analysis of a citation network. *The BMJ*, **339**, (2009), b2680.
[360] A. S. Gerber, and N. Malhotra, Publication bias in empirical sociological research: Do arbitrary significance levels distort published results? *Sociological Methods and Research*, **37**(1), (2008), 3–30.
[361] D. J. Benjamin, J. O. Berger, M. Johannesson, et al., Redefine statistical significance. *Nature Human Behaviour*, **2**(1), (2018), 6–10.
[362] O. Efthimiou, and S. T. Allison, Heroism science: Frameworks for an emerging field. *Journal of Humanistic Psychology*,. **58**(5), (2018), 556–570.
[363] B. A. Nosek, C. R. Ebersole, A. C. DeHaven, et al., The preregistration revolution. *Proceedings of the National Academy of Sciences*, **115**(11), (2018), 2600–2606.
[364] T. S. Kuhn, *The Essential Tension: Selected Studies in Scientific Tradition and Change* (Chicago: University of Chicago Press, 1977).
[365] P. Bourdieu, The specificity of the scientific field and the social conditions of the progress of reasons. *Social Science Information*, **14**(6), (1975), 19–47.
[366] L. Yao, Y. Li, S. Ghosh, et al., Health ROI as a measure of misalignment of biomedical needs and resources. *Nature Biotechnology*, **33**(8), (2015), 807–811.

[367] W. Willett, *Nutritional Epidemiology* (New York: Oxford University Press, 2012).
[368] J. M. Spector, R. S. Harrison, and M.C. Fishman, Fundamental science behind today's important medicines. *Science Translational Medicine*, 10(438), (2018), eaaq1787.
[369] A. Senior, J. Jumper, and D. Hassabis, AlphaFold: Using AI for scientific discovery. Deepmind article/blog post available online at https://bit.ly/34PXtzA (2020).
[370] Y. N. Harari, Reboot for the AI revolution. *Nature News*, 550(7676), (2017), 324–327.
[371] A. Krizhevsky, I. Sutskever, and G. E. Hinton. ImageNet classification with deep convolutional neural networks, in *Advances in Neural Information Processing Systems 25 (NIPS 2012)* (San Diego, CA: NIPS Foundation, 2012).
[372] C. Farabet, C. Couprie, L. Najman, et al., Learning hierarchical features for scene labeling. *IEEE Transactions on Pattern Analysis and Machine Intelligence*, 35(8), (2012), 1915–1929.
[373] J. J. Tompson, A. Jain, Y. LeCun, et al., Joint training of a convolutional network and a graphical model for human pose estimation, in *Advances in Neural Information Processing Systems 27 (NIPS 2014)* (San Diego, CA: NIPS Foundation, 2014).
[374] C. Szegedy, W. Liu, Y. Jia, et al., Going deeper with convolutions, in *Proceedings of the IEEE Conference on Computer Vision and Pattern Recognition* (Piscataway, NJ: IEEE, 2015), pp. 1–9.
[375] T. Mikolov, A. Deoras, D. Povey, et al., Strategies for training large scale neural network language models, in *2011 IEEE Workshop on Automatic Speech Recognition & Understanding* (Piscataway, NJ: IEEE, 2011), pp. 196–201.
[376] G. Hinton, L. Deng, D. Yu, et al., Deep neural networks for acoustic modeling in speech recognition. *IEEE Signal Processing Magazine*, 29(6), (2012), 82–97.
[377] T. N. Sainath, A. Mohamed, B. Kingsbury, et al. Deep convolutional neural networks for LVCSR, in *2013 IEEE International Conference on Acoustics, Speech and Signal Processing* (Piscataway, NJ: IEEE, 2013), pp. 8614–8618.
[378] A. Bordes, S. Chopra, and J. Weston, Question answering with subgraph embeddings. arXiv preprint https://arxiv.org/pdf/1406.3676.pdf, (2014).
[379] S. Jean, K. Cho, R. Memisevic, et al., On using very large target vocabulary for neural machine translation. arXiv preprint https://arxiv.org/abs/1412.2007, (2014).
[380] I. Sutskever, O. Vinyals, and Q. V. Le. Sequence to sequence learning with neural networks, in *Advances in Neural Information Processing Systems 27 (NIPS 2014)* (San Diego, CA: NIPS Foundation, 2014).

[381] J. Ma, R. P. Sheridan, A. Liaw, et al., Deep neural nets as a method for quantitative structure–activity relationships. *Journal of Chemical Information and Modeling,*. 55(2), (2015), 263–274.

[382] T. Ciodaro, D. Deva, J. M. de Seixas, et al., Online particle detection with neural networks based on topological calorimetry information. *Journal of Physics: Conference Series,* 368, (2012), 012030.

[383] Kaggle. Higgs boson machine learning challenge. Available online at www.kaggle.com/c/higgs-boson/overview (2014).

[384] M. Helmstaedter, K. Briggman, S. Turaga, et al., Connectomic reconstruction of the inner plexiform layer in the mouse retina. *Nature,* 500(7461), (2013), 168–174.

[385] M. K. Leung, H. Y. Xiong, L. Lee, et al., Deep learning of the tissue-regulated splicing code. *Bioinformatics,* 30(12), (2014), i121–i129.

[386] H. Y. Xiong, B. Alipanahi, L. Lee, et al., The human splicing code reveals new insights into the genetic determinants of disease. *Science,* 347(6218), (2015), 1254806.

[387] D. Silver, T. Hubert, J. Schrittwieser, et al., A general reinforcement learning algorithm that masters chess, shogi, and Go through self-play. *Science,* 362(6419), (2018), 1140–1144.

[388] J. De Fauw, J. R. Ledsam, B. Romera-Parede, et al., Clinically applicable deep learning for diagnosis and referral in retinal disease. *Nature Medicine,* 24(9), (2018), 1342–1350.

[389] A. Esteva, B. Kuprel, R. A. Novoa, et al., Dermatologist-level classification of skin cancer with deep neural networks. *Nature,* 542(7639), (2017), 115–118.

[390] J. J. Titano, M. Badgeley, J. Schefflein, et al., Automated deep-neural-network surveillance of cranial images for acute neurologic events. Nature Medicine, 24(9), (2018), 1337–1341.

[391] B. A. Nosek and T. M. Errington, Reproducibility in cancer biology: Making sense of replications. *Elife,* 6, (2017), e23383.

[392] C. F. Camerer, A. Dreber, E. Forsell, et al., Evaluating replicability of laboratory experiments in economics. *Science,* 351(6280), (2016), 1433–1436.

[393] C. F. Camerer, A. Dreber, F. Holzmeister, et al., Evaluating the replicability of social science experiments in *Nature* and *Science* between 2010 and 2015. *Nature Human Behaviour,* 2(9), (2018), 637–644.

[394] A. Chang and P. Li, Is economics research replicable? Sixty published papers from thirteen journals say "usually not." Finance and Economics Discussion Series 2015-083. Washington, DC: Board of Governors of the Federal Reserve System. Available online at https://bit.ly/34RI3uy, (2015).

[395] Y. Wu, Y. Yang, and B. Uzzi, An artificial and human intelligence approach to the replication problem in science. [Unpublished data.]

[396] M. Tegmark, *Life 3.0: Being Human in the Age of Artificial Intelligence* (New York: Alfred A. Knopf, (2017).

[397] J. Dastin, Amazon scraps secret AI recruiting tool that showed bias against women. Reuters news article, available online at https://bit.ly/3cChuwe, (October 10, 2018).

[398] Y. Wang, B. F. Jones, and D. Wang, Early-career setback and future career impact. *Nature Communications*, 10, (2019), 4331.

[399] T. Bol, M. de Vaan, and A. van de Rijt, The Matthew effect in science funding. *Proceedings of the National Academy of Sciences*, 115(19), (2018), 4887–4890.

[400] V. Calcagno, E. Demoinet, K. Gollner, et al., Flows of research manuscripts among scientific journals reveal hidden submission patterns. *Science*, 338(6110), (2012), 1065–1069.

[401] P. Azoulay, Small-team science is beautiful. *Nature*, 566(7744), (2019), 330–332.

[402] S. Haustein, I. Peters, C. R. Sugimoto, et al., Tweeting biomedicine: An analysis of tweets and citations in the biomedical literature. *Journal of the Association for Information Science and Technology*, 65(4), (2014), 656–669.

[403] T. V. Perneger, Relation between online "hit counts" and subsequent citations: Prospective study of research papers in *The BMJ*. *The BMJ*, 329(7465), (2004), 546–547.

[404] D. Li, P. Azoulay and B. N. Sampat, The applied value of public investments in biomedical research. *Science*,. 356(6333), (2017), 78–81.

[405] M. Ahmadpoor and B. F. Jones, The dual frontier: Patented inventions and prior scientific advance. *Science*,. 357(6351), (2017), 583–587.

[406] A. Duckworth and A. Duckworth, *Grit: The Power of Passion and Perseverance* (New York: Scribner, 2016).

[407] J. D. Angrist and J.-S. Pischke, The credibility revolution in empirical economics: How better research design is taking the con out of econometrics. *Journal of Economic Perspectives*, 24(2), (2010), 3–30.

[408] K. J. Boudreau, T. Brady, I. Ganguli, et al., A field experiment on search costs and the formation of scientific collaborations. *Review of Economics and Statistics*, 99(4), (2017), 565–576.

[409] H. J. Kleven, Language trends in public economics. Slides available online at https://bit.ly/2RSSTuT, (2018).

[410] C. J. Ruhm, *Deaths of Despair or Drug Problems?*, working paper 24188 (Cambridge, MA: National Bureau of Economic Research, 2018).

[411] S. Redner, How popular is your paper? An empirical study of the citation distribution. *The European Physical Journal B: Condensed Matter and Complex Systems*, 4(2), (1998), 131–134.

[412] H. Jeong, Z. Néda, and A.-L. Barabási, Measuring preferential attachment in evolving networks. *Europhysics Letters*, 61(4), (2003), 567.

[413] M. E. Newman, Clustering and preferential attachment in growing networks. *Physical Review E*, 64(2), (2001), 025102.

[414] P. L. Krapivsky and S. Redner, Organization of growing random networks. *Physical Review E*, **63**(6), (2001), 066123.

[415] G. J. Peterson, S. Pressé, and K.A. Dill, Nonuniversal power law scaling in the probability distribution of scientific citations. *Proceedings of the National Academy of Sciences*, **107**(37), (2010), 16023–16027.

[416] M. V. Simkin and V. P. Roychowdhury, Do copied citations create renowned papers? *Annals of Improbable Research*, **11**(1), (2005), 24–27.

[417] M. V. Simkin and V. P. Roychowdhury, A mathematical theory of citing. *Journal of the American Society for Information Science and Technology*, **58**(11), (2007), 1661–1673.

[418] R. A. Bentley, M.W. Hahn, and S. J. Shennan, Random drift and culture change. *Proceedings of the Royal Society of London. Series B: Biological Sciences*, **271**(1547), (2004), 1443–1450.

[419] A. Vazquez, Knowing a network by walking on it: Emergence of scaling. arXiv preprint https://arxiv.org/pdf/cond-mat/0006132v1.pdf, (2000).

[420] J. M. Kleinberg, R. Kumar, P. Raghavan, et al., The web as a graph: Measurements, models, and methods, in *Lecture Notes in Computer Science*, vol. 1627, *Computing and Combinatorics*, (Berlin: Springer-Verlag, 1999), pp. 1–17.

[421] J. P. Perdew and Y. Wang, Accurate and simple analytic representation of the electron–gas correlation energy. *Physical Review B*, **45**(23), (1992), 13244.

[422] M. V. Simkin and V. P. Roychowdhury, Read before you cite! arXiv preprint https://arxiv.org/pdf/cond-mat/0212043.pdf, (2002).

[423] J. M. Kosterlitz and D. J. Thouless, Ordering, metastability and phase transitions in two-dimensional systems. *Journal of Physics C: Solid State Physics*, **6**(7), (1973), 1181–1203.

[424] F. Radicchi and C. Castellano, A reverse engineering approach to the suppression of citation biases reveals universal properties of citation distributions. *PLoS One*, **7**(3), (2012), e33833.

INDEX

A. M. Turing Award 135
academic pipeline 167–172
   employment 168–172
   PhDs 167–168
Adam (artificial scientist) 224–226
age
   entrepreneurs 58–59
   life cycle of scientists 40
   Nobel laureates 54–58
   of scientists when most important discoveries made 5–6, 40–41
aging 210
Albert Einstein Institute 81–82
Alexandria, Library of 197, 206, 208
AlphaGo 233
altmetrics 246
Amazon 235, 239–240
American Economic Review 32
American Mathematical Society 141
American Physical Society 157
American Sociological Review 207
American Sociological Society 158
Annalen der Physik 19
articles *see* publications
artificial intelligence 232–234
   human intelligence and 237–240
   impact on science 234–240
   information organization 235–236
   limitations 239–240

narrow 238
   problem-solving 236–237
   supervised learning 233
arXiv 24
astronomy 126
ATLAS project 124
atorvastatin 123
authorship, guest authors 143
automated science 224–226
Azoulay, Pierre 251

B-29 bombers 242
Baba, Shigeru 153
Bacon, Francis 165, 198
Bardeen–Cooper–Schrieffer (BCS) paper 186, 213
basketball 71
Beard, George M. 40
Beethoven, Ludwig van 165
Bell Labs 230
Benz, Karl 190
Benz Patent-Motorwagen 189–193
Bernard of Chartres 148
Bianconi–Barabasi model 187
bibliometrics
   altmetrics 246
   citations per paper (C/N) 21
   individual performance 26
   prediction of future output 23

predictive power 23
total number of publications (N) 21
Biochemical Journal 191
biology 25–26, 97, 103, 141–142
  author order 139
Bohr, Niels 6, 17, 46, 166
Bolt, Usain 12
Boyle, Robert 99
Bragg, Lawrence 44
brain drain 93
Brazil 90
British Association for the Advancement of Science 28
Brout, R. 157
Brown, Michael S. 122
Brownian motion 71
burden of knowledge theory 45

Cardona, Manuel 24
CASP conference 231
Cattell, James McKeen 24
Cell 140, 214
cell biology 33
CERN 88, 124
chair effect 29
Chalfie, Martin 134, 157
Chambers, Paul 120
chemistry 142
  Nobel Prize 33, 80, 152–153
  productivity levels 10
chickens 111
China 90, 93
  PhD student numbers 170
citation counts
  age distribution 210–212
  prediction 215–219
citations 159
  age distribution 204
    expiration date 205–206
  benefits to author 159
  bibliographic measures based on 246
  copying 261
  counts 179
    publicity and 193–194
    "sleeping beauties" 213
  defined 159
disadvantages of metrics based on 180–182
distributions 174–177
  by age 198–199, 204–206
  lognormal 188
  universality 177–179
models
  fitness model 187–188, 210–211, 214, 260–262
  minimal model for individual papers 262
  Price model 185, 257–259
negative 181, 196
patents 183
perfunctory 181
citations per paper (C/N) 21
coauthorship networks 103
  collaborator numbers 103–106
  connected components 108
  connectedness 109
  diversity parameter 119
  incumbency parameter 119
  small world effects 106–108
  super ties 121–123
  visualized 104
  see also multi-authored papers
collaboration effects 26
collaborations 255
  coauthorship networks 103
  effects, halo effect 97
  large-scale 82
  multi-university 91
  peer effects 95–96
  prevalence 92
  too much talent 111–113
collective intelligence 117
computer science 9, 30
Computing Research Association (CRA) 10
conceptual innovators 46–49
conference papers 10
connected components 108
"constant-probability-of-success" model 55
Copernicus, Nicolaus 165

credibility revolution 249
credit allocation
  algorithm 151–158
  collective 151–158
Crick, Francis 82
criticality 128–131
Croatian Medical Journal 139
Curie, Marie 5, 53, 165
CVs 68

Dartmouth College 95
Darwin, Charles 5, 47–49
Davidson, Cathy N. 112
Davis, Miles 120
De Grasse, Andre 13
de Solla Price, Derek 161, 174, 185
Deep Blue 233
deep learning 232
DeepMind 233–234
Deng, Xiao Tie 106
digitization 202–204
Dirac, Paul 17
disruption 128–131
disruption index 130, 246
DNA 82
Drago, Marco 81–82
drug research 230
Duke University 111–113
Dylan, Bob 48

Easton, Pennsylvania 24
ecology 119–120
economics 119–120, 141, 251
  women in 144
Eddington, Arthur 20
Eddington number 20
education 167–172
Einstein, Albert 19, 82, 149
  Annus mirabilis 5, 40, 71–72
  h-index 19
  Landau greatness rank 17
  publication count 5
  quoted 5, 27, 180–182
electrospray ionization 79
employment 168–172

engineering 142
Englert, F. 157
entrepreneurs 58–59
Erdős number 102
Erdős Number Project 106
Erdős, Paul 7, 10, 14, 102, 213
ethnic diversity 114–115
"Eureka!" moments 165
experimental innovation 46–49
experimental design 248–251
exponential growth 210, 262
  significance 163–166

Facebook 235, 246
Faraday, Michael 5
fat-tailed distributions 176
Fenn, John 6, 33, 37, 58, 79
Fermi, Enrico 6
Fert, Albert 136
Fields Medal 79
file-drawer problem 228
fish oil 227
Fish, Stanley 111
fitness model 187–188, 210–211, 214, 260–262
Fleming, Alexander 6, 53
free-riding phenomenon 96
French Academy 35
Freud, Sigmund 5
Frisch, Otto 166
Frost, Robert 48
funding 36, 79, 118, 229, 247, 249
  team size and 132
future impact 215–219

Garfield, Eugene 174, 213
Gastroenterology (journal) 140
Gates, Bill 106, 108
Gates, Henry Louis Jr. 112
Gatlin, Justin 12
gender 144–146
Gene Expression 191
Genes, Brain, and Behavior 140
ghost authors 143
globalization 89

Go (game) 232
Godeaux, Lucien 102
Goldberg, Jonathan 112
Goldstein, Joseph L. 122
Goliath's shadow effect 49
Google Scholar 20, 26, 33, 202, 261
grant application databases 243
graphene 154
green fluorescent protein (GFP) 134
grit factor 248
Grossman, Jerry 106
groupthink 128
Guare, John 107
guest authors 143
Guralnik, G. S. 157

h-index 2, 17–18, 67
   advantages over other bibliometric measures 21
   collaboration effects 26
   definition 18–20
   disadvantages 25
      inter-field difference 25–26
   growth over time 19
   origin of 24
   predictive power 21–23
   time-dependence 26
Hagen, C. R. 157
Hahn, Otto 166
Harvard Medican School 249
heat 223
Heisenberg, Werner 6, 17, 46
Hell, Pavol 106
Higgs boson 157
Higgs, Peter 5, 19, 25, 157
Hirsch, Jorge E. 18, 24
history 10
Hooke, Robert 62, 159
horizon scanning 235
hot streaks 71–74
   basketball 71
   duration 77
   general human activity patterns 77
   occurrence within career 76
   Q-model and 74–78

   relative timing of hits 72
   significance 78–80
   visualized 73
Howard Hughes Medical Institute 36, 230
Human Genome Project 82
hurricane forecasting 217

i10 index 26
ImageNet challenge 232
impact 32–34
   age of citations and 199–201
   broad definition 245–248
   citation-based metrics 159–160
   future 215–219
   high-impact papers
      first-mover advantage 186–187
      fit get richer model 187–188
      rich-get-richer effect 184–186
   knowledge hotspots 199–201
   large teams and 130
   novelty and 189–193
   older discoveries 202–204
   Q-factor prediction 67–70
   relative team 86
   team diversity and 115–116
   ultimate 214–215
impact factor 119–120, 205, 207
income inequality 180
India 90
individual diversity 115
Industrial Revolution 223
information behavior 202
instrumentation 252
intelligence 117
international diversity 114
Internet Engineering Task Force (IETF) 29
inventions 189–193, 223
   impact of academic research on 247
invisible college 98–101
iPod 190
IQ 117

Jackson, Peter 75
Japan 170
jazz 120

Jobs, Steve 190
journal article *see* publications
Journal of Biological Chemistry 191, 206
Journal of Immunology 196
Journal of Personality and Social Psychology 207
journals
  impact factor 119–120, 205, 207, 216, 219, 256
  letters 166
  online access 202
Julius Caesar 197
jump-decay pattern 205, 209

Kasparov, Garry 233
Ke Jie 234
Kelly, Wynton 120
Kibble, T. W. B. 157
knowledge hotspots 199–201
Kuhn, Thomas 2

Lafayette College 24
Lancet, The 195
Landau, Lev 17
Large Hadron Collider (LHC) 88–89, 124
Lehman, Harvey 40
Lentricchia, Frank 111
letters 166
LIGO experiment 81–82, 127–128, 132
lognormal distribution 11, 176, 178, 188
low-density lipoprotein 123

machine-aided science 223
materials science 142
mathematics 85, 103, 105
  author list order 141
Matthew effect 28–29, 33, 72, 148–151, 185
  evidence for 37–38
  reverse 34, 150
media bias 195
medicine 141
Meitner, Lise 166
Mencius 96
Mendel, Gregor 5

Mermin, David 17
Merton, Robert K. 28, 97
metadata 16
Microsoft Academic Search 163
migraine headaches 227
MMR vaccination 195
molecular biology 25–26
Molecular Biology of the Cell 140
Montaigner, Luc 6
Moon, Michael 112
Moore, Gordon 172
Muir, William M. 110
multi-authored papers 16
  author contributions 138
  author lists 137–140
    alphabetical 141–144
    joint first authorship 140
    order 139
  credit assignment 26, 134–146, 147–158
    Matthew effect 148–151
    collaboration penalty for women 144–146
  retractions 150
  truncated author lists 29
  see also coauthorship networks; collaborations; teams
musicals 120

name disambiguation 16
National Academy of Sciences 22
National Institutes of Health 3, 39, 50, 99, 243, 247
  awards 79
  principal investigators 39
National Research Council 10
National Science Foundation 3, 82, 118, 235
  Survey of Doctorate Recipients 171
National Science Fund for Distinguished Young Scholars 79
natural experiments 250
Nature 31, 137, 166, 204, 207
  doctoral student survey 168
  joint first author papers 140
  Science Citation Index tally 174

negative citations 181, 196
Negishi, Ei-ichi 153
neural networks 232–234
neuroscience 142
New England Journal of Medicine 193
New York Times 112, 193
Newton, Isaac 17, 35, 159
   annus mirabilis 40
   quoted 45, 62, 198
Nobel Prize
   age of winners 44, 57
   attention accruing to winners' previous work 36
   banquet ceremony (2008) 134
   career patterns and 57
   chemistry 33, 80, 152–153
   delay in award 45
   medicine 123
   physics 44, 136, 151–152, 157
   prediction using Q-factor 67–70
   winners lost to mortality 57
novelty
   measurement 189–193
   uncertainty and 191

Oakland University 106
Occupy Wall Street 180
Olympic Games 12–13
orphan enzymes 225

Papadimitriou, Christos 106
paradigm shifts 2
Pasteur, Louis 5
patent applications 88, 243, 247
   citations 183
   number of inventors per applications 125
Patterson, Annabel and Lee 112
peer effects 95–96
   causal inference for 98
   interaction and 99
peer review 30–32, 204
Perdew, John P. 261
PhD programs 10, 167–168
   thesis length 168

Philosophical Transactions of the Royal Society 161
photoelectric effect 71
Physical Review 209
Physical Review B 207, 216, 261
Physical Review Letters 9, 19, 53, 124, 136, 166, 180–181, 207
physics 33, 67–70
   1905 as miracle year 71–72
   multi-author papers 139, 141–142, 151–158
   Nobel Prize 44, 136, 151–152, 157
   publication volumes 162
Picasso, Pablo 48, 165
Planck, Max 49
Planck's principle 49
Poisson, Siméon Denis 5
postdoctoral fellowships 169, 171
power-law distributions 77, 126, 176
Prasher, Douglas 134, 157
preferential attachment 105, 187, 210, 257–258, 261–262
   origins of 259–260
preregistration 229
prizes 229
   bias toward individual work 135
Proceedings of the National Academy of Sciences 216
productivity 13–15, 63
   contributing factors 13–15
   economic definition 172
   eminence and 15
   funding and 248–251
   hot streaks and 77
   impact 54
   interdisciplinary comparison 8–11
   name identification and 16
   peer effects 99
   R&D workers 45
   typical scientist 10
prospective citations 204
protein folding 231
Prussian Academy of Science 82

psychology 119–120, 128
  author lists 139
  article length 8–9
  cognitive space and 173
  exponential growth 8, 162–163
    significance 163–166
  total 163
  visualized 175
publications, manuscript rejections 245
publicity 193–194
Purdue University 9, 110

Q-model 61–64
  age and experience 66
  calculation 65–66
  consistency 64
  hot streaks and 74–78
  impact prediction 67–70
quasi experiments 250

R-model 62
Radway, Janice 112
random impact rule 53–58
randomized controlled trials (RCT) 249–250
Rayleigh, Lord 28, 148
receiver operating characteristic (ROC) 67
Rényi, Alfréd 213
replication 237–240
reproducibility crisis 237
Reproducibility Project 237
reputation 32
retractions 34, 150
retrospective citations 204
Rich, Barnaby 166
rich-get-richer effect 184–186, 258–260
risk management 230
risk-taking 229
robot scientists 224–226
Royal Society 161
Rubbia, Carlo 151, 153
Rutherford, Ernest 6, 165–166

Sakurai prize 157
SARA method 26

ScholarOne 235
Schrödinger, Erwin 17
Science Citation Index 174
Science (journal) 33, 140
science of science 1
science of team science (SciTS) 83
scientific impact *see* impact
scientists
  life cycle 40
  numbers of 9
second acts 213
Sedgwick, Eve Kosofsky 112
selection effect 98, 196
Shockley, William 11
shogi (Japanese chess) 232
significant papers 21
Six Degrees of Separation 107
sleeping beauties 213
small world phenomenon 106–108
Smith, Barbara Herrnstein 112
social media 246
social sciences 85, 139
Sommerfeld, Arnold 46
Soviet Union 99
special relativity 71
Stanford University 122
star researchers 97, 100
  too-much-talent effect 111–113
steam engine 223
super ties 121–123
superconductivity 186, 213
superfluidity 17
supervised learning 233
survivorship bias 241–245
Swanson hypothesis 227

teams 249
  benefits to individuals 88, 93, 95
  collective intelligence 117
  composition 118–121, 254
    balance 113–118
    ethnic diversity 114–115
    international diversity 115
    size and 125–127
    too-much-talent effect 111–113
  dominance of science 86–88

drivers behind 88–89
formation process 126, 255
idea generation 86
invisible college 98–101
large 124
longevity 127
modeling 254
multi-university collaborations 91
productivity 110
relative impact 86
size 127–128
  disadvantages of large 128
  disruption vs. development 128–131
  small 132
small world effect 103–106
specialization and 88
transnational collaboration 89–94
TechCrunch awards 58
technicians 143
thermodynamics 223
Thomas, Dorothy Swaine 147, 149, 158
Thomas theorem 147
Thomas, W. I. 147, 149
Tompkins, Jane 112
too-much-talent effect 111–113
total number of publications (N) 21
tradition and risk 229
Tsien, Roger 134

ultimate impact 214–215
  trajectories toward 218
United States 167
  PhDs 170
US News and World Report 10, 112

vaccinations 195
van de Rijt, Arnout 37

van der Meer, Simon 151, 153
van Gogh, Vincent 75
video games 120
Virginia Commonwealth University 79

W and Z bosons 151–158
Wakefield, Andrew 195
Wald, Abraham 242
Warhol, Andy 48
Watson, James 82
weather forecasts 215, 217
Web 2.0 246
Web of Science 10, 131, 162, 200
  paper mountain 175
Web Search and Data Mining (WSDM) conference 30
white spaces 227–230
Wigner, Eugene 6
Wilczek, Frank 238–239
Wilson, Kenneth G. 52
Wineland, David J. 136
women
  algorithmic discrimination against 239–240
  collaboration penalty 144–146
Woods, Tiger 13
World War I and II 42, 251
Wright, Frank Lloyd 48

Yale University 79
yeast 224
Yongle Encyclopedia 87

Z boson 151–158
Zallen, Dick 24
Zerhouni, Elias 39, 50
Zuckerman, Harriet 88, 148